# Assessment of Arsenic Treatment Residuals

# Analysis and Stabilization Techniques

AF849281

## About the Awwa Research Foundation

The Awwa Research Foundation (AwwaRF) is a member-supported, international, nonprofit organization that sponsors research to enable water utilities, public health agencies, and other professionals to provide safe and affordable drinking water to consumers.

The Foundation's mission is to advance the science of water to improve the quality of life. To achieve this mission, the Foundation sponsors studies on all aspects of drinking water, including supply and resources, treatment, monitoring and analysis, distribution, management, and health effects. Funding for research is provided primarily by subscription payments from approximately 1,000 utilities, consulting firms, and manufacturers in North America and abroad. Additional funding comes from collaborative partnerships with other national and international organizations, allowing for resources to be leveraged, expertise to be shared, and broad-based knowledge to be developed and disseminated. Government funding serves as a third source of research dollars.

From its headquarters in Denver, Colorado, the Foundation's staff directs and supports the efforts of more than 800 volunteers who serve on the board of trustees and various committees. These volunteers represent many facets of the water industry, and contribute their expertise to select and monitor research studies that benefit the entire drinking water community.

The results of research are disseminated through a number of channels, including reports, the Web site, conferences, and periodicals.

For subscribers, the Foundation serves as a cooperative program in which water suppliers unite to pool their resources. By applying Foundation research findings, these water suppliers can save substantial costs and stay on the leading edge of drinking water science and technology. Since its inception, AwwaRF has supplied the water community with more than $300 million in applied research.

More information about the Foundation and how to become a subscriber is available on the Web at **www.awwarf.org.**

# Assessment of Arsenic Treatment Residuals

# Analysis and Stabilization Techniques

Prepared by:
**Timothy A. Kramer**, **Richard H. Loeppert**, and **Hunyoung Wee**
Texas A&M University, College Station, TX 77843

Jointly sponsored by:
**Awwa Research Foundation**
6666 West Quincy Avenue, Denver, CO 80235-3098

**U.S Environmental Protection Agency**
Washington, DC

and

**Association of California Water Agencies**
Sacramento, CA

Published by:

Awwa Research Foundation

Distributed by:

American Water Works Association

IWA Publishing

## DISCLAIMER

This study was funded under the Arsenic Research Partnership. The Arsenic Research Partnership consisted of the Awwa Research Foundation (AwwaRF), the U.S. Environmental Protection Agency (USEPA), and the Association of California Water Agencies (ACWA) under Cooperative Agreement No. CR826432-01. AwwaRF, USEPA, and ACWA assume no responsibility for the content of the research study reported in this publication or for the opinions or statements of fact expressed in the report. The mention of trade names for commercial products does not represent or imply the approval or endorsement of AwwaRF, USEPA, or ACWA. This report is presented solely for informational purposes.



ISBN 978-1-84339-215-6

# CONTENTS

# LIST OF TABLES

# LIST OF FIGURES

# FOREWORD

The Awwa Research Foundation is a nonprofit corporation that is dedicated to the implementation of a research effort to help utilities respond to regulatory requirements and traditional high-priority concerns of the industry. The research agenda is developed through a process of consultation with subscribers and drinking water professionals. Under the umbrella of a Strategic Research Plan, the Research Advisory Council prioritizes the suggested projects based upon current and future needs, applicability, and past work; the recommendations are forwarded to the Board of Trustees for final selection. The foundation also sponsors research projects through the unsolicited proposal process; the Collaborative Research, Research Applications, and Tailored Collaboration programs; and various joint research efforts with organizations such as the U.S. Environmental Protection Agency, the U.S. Bureau of Reclamation, and the Association of California Water Agencies.

This publication is a result of one of these sponsored studies, and it is hoped that its findings will be applied in communities throughout the world. The following report serves not only as a means of communicating the results of the water industry's centralized research program but also as a tool to enlist the further support of the nonmember utilities and individuals.

Projects are managed closely from their inception to the final report by the foundation's staff and large cadre of volunteers who willingly contribute their time and expertise. The foundation serves a planning and management function and awards contracts to other institutions such as water utilities, universities, and engineering films. The funding for this research effort comes primarily from the Subscription Program, through which water utilities subscribe to the research program and make an annual payment proportionate to the volume of water they deliver and consultants and manufacturers subscribe based on their annual billings. The program offers a cost-effective and fair method for funding research in the public interest.

A broad spectrum of water supply issues is addressed by the foundation's research agenda: resources, treatment and operations, distribution and storage, water quality and analysis, toxicology, economics, and management. The ultimate purpose of the coordinated effort is to assist water suppliers to provide the highest possible quality of water economically and reliably. The true benefits are realized when the results are implemented at the utility level. The foundation's trustees are pleased to offer this publication as a contribution toward that end.

David E. Rager
Chair, Board of Trustees
Awwa Research Foundation

Robert C. Renner, P.E.
Executive Director
Awwa Research Foundation

## ACKNOWLEDGMENTS

Hun-Young Wee and Yoko Masue were graduate students at Texas A&M University who conducted the experiments of this project and analyzed the data from them. This project would not have made any progress without them.

The participating utilities provided water treatment plant residuals which were essential to the completion of this project and their contributions are gratefully acknowledged. They include the El Paso Water Utilities (El Paso, TX), Public Utilities Department of the City of Billings (Billings, MT), and the Helena Water Treatment Utilities (Helena, MT), and Naval Air Station Fallon (Fallon, NV). CH2M-HILL also cooperated by providing residuals from a pilot plant they operated for the El Paso Water Utilities.

The Project Advisory Committee (PAC) provided their knowledge and judgment throughout the project, and their contributions are appreciated. The PAC consisted of Camille Brown (United Water Idaho), Mike MacPhee (Malcolm Pirnie), John Novak (Virginia Polytechnic Institute and State University), and Ram Ramaswami (Environmental Resources Management).

The AwwaRF Project Manager was Alice Fulmer, and her patience and persistence were critical in the completion of this project.

The authors gratefully acknowledge the contributions of Dr. Bill Batchelor in editing the final report.

## IN MEMORIAM

During the preparation of this final report, Dr. Timothy A. Kramer passed away. This report is dedicated to the memory of Dr. Kramer and his life of service to his students, friends, and profession. He was an outstanding teacher, gifted mentor and advisor, and accomplished researcher.

# EXECUTIVE SUMMARY

Arsenic is a toxic compound with a complex aqueous chemistry. The USEPA has mandated a reduction in the maximum contaminant level for arsenic from 50 ppb to 10 ppb effective in 2006. This more stringent removal criterion and the resulting increased need to remove arsenic from water will lead to the production of large volumes of arsenic-tainted residuals. A common method of removing arsenic from contaminated water is by the co-precipitation or sorption of the arsenic onto the oxy-hydroxides formed from the addition of various metal salts. Two such metal salts of interest are aluminum sulfate and ferric chloride. This form of treatment produces solids containing high levels of arsenic, which, in common practice, would be disposed in landfills. However, the elevated arsenic content could pose problems with leaching in the natural environment, thus requiring expensive disposal in certified hazardous waste impoundments. In addition to co-precipitation using metal salts, several proprietary granular media have been produced to adsorb arsenic in packed bed, flow-through columns. As with the metal salts, the granular media, with or without media regeneration, pose potential disposal problems.

The focus of this project concerns establishing the proper methods of post-treatment for arsenic-containing residuals, leading to safe and inexpensive disposal. The objectives of this research are three-fold. First, a general physical and chemical characterization will be made of arsenic-contaminated residuals from water treatment processes. This project phase included not only well characterized laboratory produced arsenic-treatment residuals, but utility and pilot study specimens as well. Second, a thorough understanding of factors impacting the release of arsenic from the residuals will be established. A concise understanding of this release behavior is needed to properly dispose of the residuals produced during arsenic-removal processes. The ability to predict the release behavior for a given chemistry will be developed. Finally, low cost methods to aggressively bind the arsenic to a variety of residuals will be evaluated.

## SORPTION AND DESORPTION STUDIES

Residuals were obtained by the removal of arsenite (As(III)) or arsenate (As(V)) from pure solutions or in the presence of kaolinite (a 1:1 clay mineral), montmorillonite (a 2:1 clay mineral), and organic material (100 mg/L carboxymethylcellulose and 100 mg/L of humic acid). Production of residual was achieved by co-precipitation/sorption using aluminum sulfate (alum), ferric chloride and lime/soda addition. Ferric chloride or alum was added at a 20:1 molar ratio of Fe or Al to As to a solution containing either arsenate or arsenite at a concentration of 100 mg/L. Calcium carbonate was also used as a sorbent for arsenic; however only half of the arsenic concentration as that utilized in the other formulations was used. The extent of arsenite and arsenate removal for each sample composition was quantified, and the residuals were analyzed for iron, aluminum and calcium contents. Desorption studies were also conducted by means of TCLP, pH variance (4, 6, 8 and 10), and the presence of competing anions (chloride, sulfate and phosphate). Kinetic studies of desorption under the conditions cited were also conducted, and the buffering capacities of the residuals documented. The specific treatments were:

1) Ferric chloride and arsenate;
2) Ferric chloride, arsenate and kaolinite;
3) Ferric chloride, arsenate and montmorillonite;

4) Ferric chloride, arsenate, carboxymethylcellulose and humic acid;
5) Ferric chloride and arsenite;
6) Ferric chloride, arsenite and kaolinite;
7) Ferric chloride, arsenite and montmorillonite;
8) Ferric chloride, arsenite, carboxymethylcellulose and humic acid;
9) Alum and arsenate;
10) Alum, arsenate and kaolinite;
11) Alum, arsenate and montmorillonite;
12) Alum, arsenate, carboxymethylcellulose and humic acid;
13) Alum and arsenite;
14) Alum, arsenite and kaolinite;
15) Alum, arsenite and montmorillonite;
16) Alum, arsenite, carboxymethylcellulose and humic acid;
17) Calcium carbonate and arsenate;
18) Calcium carbonate and arsenite.

Ferric chloride was much better at removing both arsenate and arsenite, whether in pure solution or in the presence of colloidal phases. Interestingly, the colloidal phases appeared to have a minimal effect on removal efficiency. It was expected prior to conducting these tests that the 2:1 clay mineral, montmorillonite, would have a positive influence, but the data indicated that, if anything, its presence resulted in less removal of both arsenate and arsenite.

TCLP extraction kinetics data were obtained using the leachate fluid #1 as described in Appendix B. Very little arsenate and arsenite were extracted from the iron-based residuals. However, both the alum and calcium carbonate based residuals showed high arsenite desorption with the TCLP protocol. As pH was increased, arsenite desorption generally decreased and arsenate desorption generally increased. It is clear that the iron-based residuals showed minimal release of both arsenate and arsenite, while the alum residuals released large amounts of arsenite at all pH levels. Desorption in the presence of phosphate showed that the ferric chloride residuals released large amounts of arsenate in the presence of phosphate. Arsenic desorption from the ferric chloride residuals upon treatment with phosphate was substantial and the elimination of this adverse behavior was a subsequent research goal. In this portion of the study, the following conclusions were drawn:

1) Iron oxy-hydroxides are far superior at removing both arsenate and arsenite, as opposed to lime/soda softening agents or aluminum oxy-hydroxides.
2) Desorption of arsenic from the iron oxy-hydroxides was minimal for both oxidation states of arsenic. The amount of arsenate desorbed was below detection limits at pH less than or equal to 8 for the iron residuals.
3) Arsenite residuals formed using alum showed by far the most arsenic desorption when compared to calcium carbonate or iron oxy-hydroxides.
4) While both aluminum and iron showed increased desorption of arsenate at higher pH, calcium carbonate showed reduced desorption under these conditions. Thus, a mixed calcium and iron system may be desirable.
5) In general, high pH favored arsenate desorption while low pH favored arsenite release (except for the calcium system).

6) Each of the residuals exhibited pH buffering characteristics. The iron and aluminum residuals tended to maintain a neutral pH while the calcium residuals raised pH.
7) Phosphate significantly enhanced desorption from iron based residuals; sulfate less so and chloride had a negligible effect. Calcium based residuals also showed a competing ion influence, while arsenic was desorbed from the aluminum residuals to a very high extent regardless of the presence of competing ions.

This portion of the investigation confirmed that arsenic can be readily removed from contaminated water using relatively simple techniques. These techniques involved the sorption of arsenate or arsenite onto the surface of iron or aluminum oxy-hydroxides. In addition, calcium carbonate was shown in this investigation to remove arsenic to acceptable levels. All of these processes are simple, robust, and easily incorporated into conventional or more advanced water treatment systems. However, they all produce residuals that can, under the appropriate circumstances, produce an environmental problem upon disposal. For example, this and other research has shown that the addition of ferric chloride ($FeCl_3$) to waters with a wide range of compositions has consistently resulted in removal of arsenic to levels well below the MCL established by the USEPA. However, when the residuals were exposed to low levels of phosphate (1.3 mM) or a pH greater than 8, substantial arsenic was released and could contaminate the environment if improperly disposed.

## BUFFERING CAPACITY OF THE ARSENIC RESIDUALS

A further phase of the investigation focused on establishing a proper measure of the buffering capacity of the residuals. Buffering capacity of arsenic residuals, or any residuals formed by the addition of metal salts, is usually substantial. If the standard TCLP test is conducted without paying particular attention to the solution pH, what might occur is not leaching at the desired TCLP pH of 4.93, but at a different pH value due to the buffering abilities of these residuals. Exactly how to quantify this ability (or capacity) of residuals to buffer water was investigated, and the results are highlighted in Chapter 3. Briefly, a standard titration can be conducted on a residuals sample of known solids content. The titration is conducted by adding small amounts of either acid or base to a one liter mixture and monitoring and recording pH. A plot is made of the amount of acid or base added versus the resulting pH. By determining the slope of the resulting titration curve and plotting the slope against pH, the buffering capacity can be obtained. This capacity is given by the following equation:

$$\beta = \frac{\partial C_B}{\partial pH} = -\frac{\partial C_A}{\partial pH} \qquad \text{(ES.1)}$$

The buffering capacity β (equivalent/L) can also be calculated using the surface acidity equilibrium constants that are also obtained from the titration data. These computations require knowledge of the surface properties of the residuals, which are often quite difficult to obtain. By knowing β, the aqueous solution pH can be computed for any given amount of acid or base added ( $\Delta C_B$ or $\Delta C_A$ )

## DESORPTION INHIBITION USING CALCIUM ADDITION DURING COAGULATION

At the initiation of this project the effort focused on obtaining arsenic treatment residuals of a known composition for both arsenate and arsenite removal. In the production of these residuals sodium hydroxide (NaOH) was used to control pH. It was noted that with the use of ferric chloride ($FeCl_3$), arsenate leaching was minimal under most conditions ($4 < pH < 9$, TCLP extraction, etc.). Thus, to minimize the arsenic residuals leaching problem, the best way to remove arsenic from contaminated drinking water would involve the oxidation of the arsenic to arsenate and the use of $Fe^{3+}$ as the sorptive oxy-hydroxide. Unfortunately, when arsenate/Fe residuals samples were tested for leaching in the presence of phosphate (1.3 mM), increased leaching was found. The importance of this result is not that arsenic is likely to be released by contact with high concentrations of phosphate in the environment, but rather that there are conditions under which arsenic will be released, and these should be mitigated to the extent possible. A method to mitigate this effect was identified when it was noted that the calcium carbonate residuals resisted leaching in the phosphate environment and that this behavior was found to be consistent with the literature. A hypothesis was developed that during coagulation the addition of calcium as calcium hydroxide ($Ca(OH)_2$) to control pH would be preferred to the use of NaOH. Excellent results were obtained and minimal leaching was found from the arsenate/$Fe^{3+}$/$Ca^{2+}$ residuals during to phosphate competition.

Desorption versus pH results indicated that the Fe/Ca system is much better than the Fe/Na mixture in retaining arsenate. Initially there is a large release of arsenate in the Fe/Na mixture, followed by some re-adsorption. The addition of calcium almost eliminated arsenate desorption in the presence of phosphate. The investigators offer at two hypotheses for this behavior. One is that the calcium is forming solid phase calcium phosphate and the other is that the calcium is altering the surface chemistry of the iron oxy-hydroxide.

## ANALYSIS OF FIELD SAMPLES

A further goal of this study was to gather characteristic samples of arsenic-containing residuals from various treatment works. Samples were obtained from three water treatment plants and two pilot studies. These residuals were selected as they were representative of the common arsenic residuals that can be expected at water treatment plants. Specifically, two alum co-precipitated residuals, one ferric chloride sample, activated alumina, a granular ferric hydroxide and a granular ferric oxide were obtained. Digestion and subsequent elemental analysis was conducted on each of the residual samples. As expected, the granular iron-based sorbents had very high levels of iron as well as arsenic. The granular alumina sorbent did not have as much arsenic as the iron-based sorbents.

The amount of arsenic leached from water treatment residuals by the toxicity characteristic leaching test (TCLP) was relatively low due to the pH of the leachate, which is favorable for As(V) adsorption. The pH values of the leachate for TCLP extraction solutions #1 and #2 were 5.1 to 6.6 and 3.5 to 4.9, respectively, depending on the residual type. The released arsenic concentrations by TCLP #2 were generally higher than those by TCLP #1.

The presence of phosphate could produce an unfavorable condition for the binding of arsenic onto the adsorbents. The extracted As(V) concentrations by phosphate competition were substantial for each of the water treatment residuals. The effects of sulfate and chloride were negligible.

The pH value of the leachate was the most important factor controlling arsenic behavior in the residuals. High concentrations of arsenic release were detected at low and high pH due to the increase in solubility of adsorbents such as Fe and Al hydroxides and the predominant arsenic species. The results showed that maintaining neutral pH in the environment of the residuals and minimizing the influent of arsenic-competing ions such as phosphate are very important to enhance the stability of arsenic when arsenic-containing residuals are disposed in a landfill without post-treatment.

## STABILIZATION TECHNIQUES

Stabilization experiments were conducted on several representative samples. Due to the extremely large number of repetitive tests two residuals were selected for study: the granular ferric hydroxide and the activated alumina. These two were chosen based on the combination of high arsenic content, iron and aluminum based metal oxy-hydroxides and a large quantity being available of each type. Two extractions were finally selected for study after it was noted that the other environmental cases did not produce appreciable arsenic leaching. The TCLP test # 2 was utilized as it was found to be the worst case for arsenic leaching of the two TCLP protocols. The extreme condition was found to be a leachate solution containing 0.1 M phosphate at a pH of 10.

The additions of lime (CaO), fly ash, ferrous sulfate, and ordinary Portland cement (OPC) were examined to determine if these compounds are suitable binders for solidification and stabilization. Only lime and OPC demonstrated a significant influence on minimizing arsenic leaching. Lime had the greatest impact on reducing the leaching of arsenic from the treatment residuals. The addition of excess lime eliminated arsenic leaching using the high pH phosphate leachate. OPC had a minimal influence on the leaching of arsenic under the worse case conditions. However, the investigators believe that the excess dissolved calcium needed to maintain the stability of the residuals may be lost over long periods of time due to its reaction with $CO_2$ from the atmosphere.

The investigators formulated several hypotheses as to why the addition of excess calcium would prove to be so influential in the elimination of arsenic leaching under all conditions tested. These hypotheses were (1) solid phase calcium phosphate compounds are formed, thus removing phosphate from competing for sorption sites on the metal oxy-hydroxides, (2) the calcium improves surface charge properties of the metal oxy-hydroxide and is conducive to conditions favorable for arsenate adsorption, and, (3) solid-phase arsenic and calcium compounds are formed. The reduction of arsenic concentration in the leachate can be obtained by the addition of lime to the residuals due to the formation of less soluble and more stable calcium arsenic compounds. In practical aspects, solidified materials treated only by lime are not stable for the long term since the buffering effects of lime are decreased by dissolution and carbonation when exposed to atmospheric $CO_2$. Therefore, it is suggested that cement should be added with the lime in order to prevent and minimize the arsenic leaching over extended periods of time.

The immobilization of arsenic in the residuals can be achieved through the formation of less soluble calcium arsenic compounds using lime and cement. Therefore, solidification and stabilization (S/S) techniques with lime and cement can be successfully applied to the immobilization of elevated arsenic concentrations in water treatment residuals.

# CHAPTER 1: PROJECT INTRODUCTION

## OVERVIEW AND PROJECT SIGNIFICANCE

Arsenic is commonly recognized as a toxic and carcinogenic compound (Bates et al., 1992; Smith et al., 1992) with well documented adverse health effects to humans (USEPA, 2000). Natural sources of arsenic are prevalent in the environment. In addition, human activities generate anthropogenic arsenic compounds (Ferguson and Gavis, 1972). Riveros et al. (2001) rigorously reviewed the arsenic generation problems and disposal practices of the mining and metallurgical industries. Natural water becomes contaminated by the discharge of arsenic through geochemical processes: dissolution of arsenic contaminated iron oxy-hydroxides under reducing conditions, oxidation of arsenic-bearing minerals, and desorption of arsenic by a competing ligand, e.g., phosphate (Bose and Sharma, 2002; Mariner and Willey, 1976).

Arsenic contamination of groundwater and/or drinking water has been reported as a critical water quality issue in Vietnam, Bangladesh, and West Bengal, India (Berg et al., 2001; Chakraborti et al., 1999; Nickson et al., 1998). In particular, 70 million people in Bangladesh are at risk due to naturally occurring arsenic in wells used for drinking water (Lepkowski, 1998). Highly elevated concentrations of arsenic (greater than 1 mg/L) have been detected in the area. Recently, more attention has been paid to arsenic removal in water treatment due to the U.S. Environmental Protection Agency (USEPA) lowering of the maximum contaminant level (MCL) from 50 to 10 mg/L, effective in 2006 (USEPA, 2001). Therefore, it is expected that many water treatment plants (WTPs) may require additional treatment technologies or modification of existing treatment technologies to meet the revised arsenic MCL. Precipitation/co-precipitation with ferric salts is considered by U.S. EPA to be the "best demonstrated available technology" (BDAT) for the removal of arsenic in water and has been successfully applied to treat arsenic in groundwater, surface water, mine drainage, drinking water, and industrial wastewater (USEPA, 2002). To meet the revised MCL for arsenic, additional or retrofit treatment may be required at many water treatment plants. Thus, appreciable volumes of arsenic-contaminated residuals are expected to be produced, which may require further treatment and handling prior to ultimate disposal.

Land filling or surface application has been used to dispose of water treatment residuals. Prior to this disposal, the residuals must meet certain leaching protocols. One measure of the release of arsenic is the USEPA TCLP test in which a sample of the residuals is exposed to water at a pH of approximately 5.0. However, at a pH of 5.0 the samples are in an environment that is more favorable to sorption rather than leaching of arsenate. Other aqueous conditions might be more favorable to leaching than what is ascertained by the approved TCLP method. Further steps might be required to effectively bind arsenic under worst-case environmental exposure, to render the residuals safe for ultimate disposal.

## OBJECTIVES

The focus of this project concerns establishing the proper methods of post-treatment for arsenic-containing residuals, leading to safe and inexpensive disposal. The objectives of this research are three-fold. First, a general physical and chemical characterization will be made of arsenic-contaminated residuals from water treatment processes. This will include not only well characterized laboratory produced arsenic treatment residuals, but utility and pilot study specimens as well. Second, a thorough understanding of the release of arsenic in natural environments will be established. A concise understanding of this release behavior is needed to properly dispose of the residuals produced during arsenic removal processes. The ability to predict the release behavior for a given chemistry will be developed. Finally, low cost methods to aggressively bind the arsenic to a variety of residuals will be obtained.

The first step in the investigation concerned desorption of arsenate and arsenite from a variety of synthesized residuals. Step two established the conditions for possible arsenic leaching. Specifically, pH and the influence of competing ions were explored. In the third step, the influence of pH and competing ions on field samples were evaluated. Samples of arsenic tainted residuals were obtained for this phase of the study and analyzed for their chemical and physical composition. Finally, techniques to non-reversibly bind the arsenic to a variety of residuals were examined.

## AQUATIC CHEMISTRY OF ARSENIC

Arsenic is considered to be a nonmetal (sometimes referred to as a metalloid) with an atomic number of 33 and an atomic mass of 74.922 grams per mol. The chemistry of arsenic is extremely complicated due to its stability in four different oxidation states, +5, +3, 0, and –3, which are dependent upon redox conditions (Gupta and Chen, 1978). However, in most natural surface waters, the primary oxidation states are +5 and +3. Arsenic species with a +5 oxidation state are known as arsenate or arsenic acid, whereas species with a +3 oxidation state are referred to as arsenite or arsenous acid (Masscheleyn et al., 1992). In addition to these inorganic forms, organic compounds containing arsenic can also be found in water. Figure 1.1 shows the chemical formulas for some of these compounds (Smith, 1973).

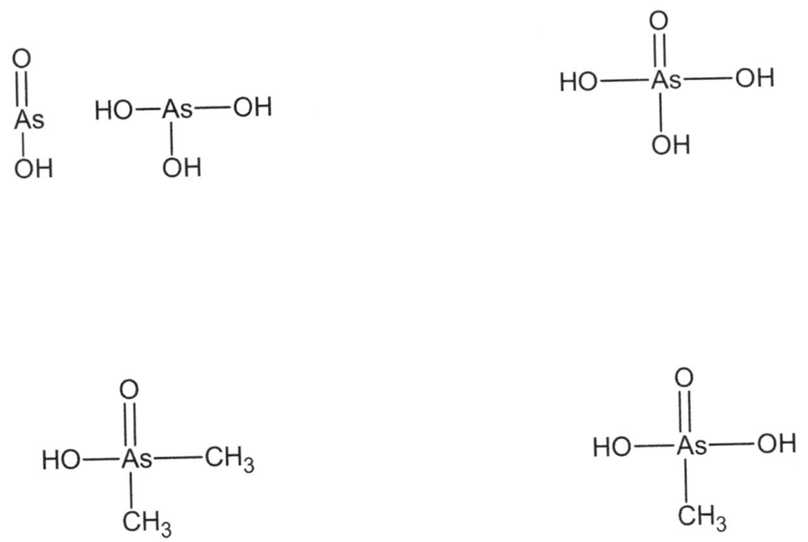

Source: Smith, 1973

**Figure 1.1 Arsenic Species in Water**

The common and most stable forms of arsenic occurring in oxidized surface waters are the arsenate species, $H_3AsO_4$, $H_2AsO_4^-$, and $HAsO_4^{-2}$. The arsenite species, $H_3AsO_3$, $H_2AsO_3^-$, and $HAsO_3^{-2}$, are three to twenty times more toxic than the arsenate species (Frank and Clifford, 1986) and are predominant under moderately reducing conditions (Gupta and Chen, 1978).

The following acid/base reactions are for the arsenate species:

$$H_3AsO_4^0 \leftrightarrow H_2AsO_4^- + H^+ \qquad pK_{a1} = 2.20 \qquad (1.1)$$

$$H_2AsO_4^- \leftrightarrow HAsO_4^{-2} + H^+ \qquad pK_{a2} = 6.97 \qquad (1.2)$$

$$HAsO_4^{-2} \leftrightarrow AsO_4^{-3} + H^+ \qquad pK_{a3} = 11.53 \qquad (1.3)$$

While arsenite exhibits these acid/base reactions:

$$H_3AsO_3^0 \leftrightarrow H_2AsO_3^- + H^+ \qquad pK_{a1} = 9.22 \qquad (1.4)$$

$$H_2AsO_3^- \leftrightarrow HAsO_3^{-2} + H^+ \qquad pK_{a2} = 12.13 \qquad (1.5)$$

$$HAsO_3^{-2} \leftrightarrow AsO_3^{-3} + H^+ \qquad pK_{a3} = 13.40 \qquad (1.6)$$

Given the above reactions, it is apparent that arsenate species are present in the form of an anion at the normal pH range (6-9) of natural waters. In contrast, the predominant arsenite species are neutrally neutral charged in this range. The neutral charge renders the arsenite species more mobile and less adsorbable than the arsenate species (Masssheleyn et al., 1992); an important characteristic when considering removal of arsenic by adsorption. This speciation has substantial implications to arsenic removal and possible desorption from residuals, as will be shown.

The distribution and mobility of dissolved arsenic species are dependent on pH and redox potential, which can be measured as Eh (volt) or pE (dimensionless). The pH and Eh relationship is important in understanding arsenic removal from water, arsenic immobilization/stabilization on solid phases as well as the distribution of arsenic species in water. Figure 1.2 illustrates the effect of pH and Eh (or pE) on major arsenic species at equilibrium conditions (Smedley and Kinniburgh, 2002). As shown in Figure 1.2, As(V) species are dominant under oxidizing conditions and As(III) is thermodynamically stable under mildly reducing conditions (Masscheleyn et al., 1991; Cherry et al., 1979). Normally, the Eh values of surface water and ground water are high and low, respectively. Thus, As(V) is more likely to occur in surface waters and As(III) tends to occur more frequently in ground waters.

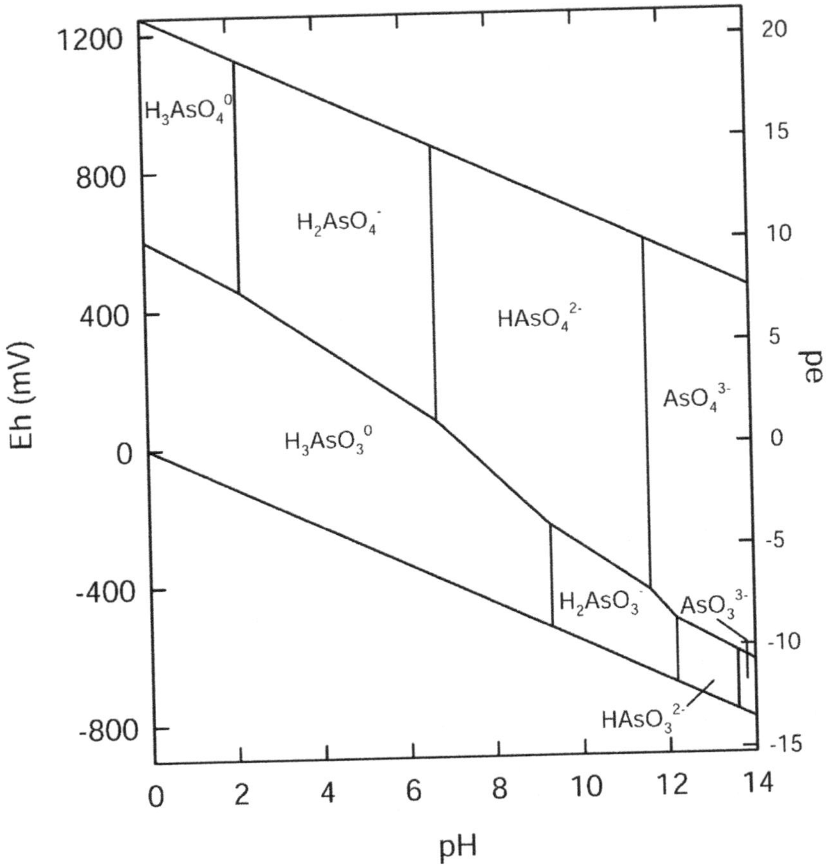


Source: Smedley and Kinniburgh, 2002.

**Figure 1.2 – Eh/pH Diagram for the As-$H_2O$ System at 25°C**

As(III) is more mobile because it is present as a neutrally charged species at the pH of most natural environments ( < pH 9) and it is less strongly adsorbed on mineral surfaces (Korte and Fernando, 1991). Meng et al. (2001) pointed out that three redox zones can be divided according to arsenic mobility. They identified "an adsorption zone" at Eh > 0, "a mobilization (transition) zone" at -0.24 < Eh (volt) < 0, and "a reductive fixation zone" at Eh (volt) < -0.24. Arsenic mobility is greatest in the mobilization zone due to the reduction of ferric oxy-hydroxides, the main adsorbents for arsenic, to ferrous iron and As(V) to As(III). Three redox zones (oxidizing, mildly reducing, and very reducing) for arsenic were also suggested by several other investigators (Carbonell-Barrachina et al., 1999; Masscheleyn et al., 1991). Figure 1.2 is useful in understanding the complex redox reactions of arsenic, but it does not consider other important factors. For example, the effects of coexisting solutes such as iron, aluminum, phosphates, silicates, and carbonates, are not considered. Furthermore, the existence of organic forms of arsenic is not considered.

Several field investigations demonstrated that the pE/pH diagram for arsenic was simplistic and that behavior was not accurately described (Williams et al., 1996; Korte and Fernando, 1991; Cullen and Reimer, 1989). As(III) and As(V) have been detected in both oxidizing (e.g., surface water) and reducing (e.g., groundwater) environments. Several researchers explain that the kinetic control of redox reactions might result in this disagreement (Peterson and Carpenter, 1983; Andreae, 1981). As(III) may be precipitated with sulfides at low

Eh, and iron oxy-hydroxides would remove both As(III) and As(V) in oxidizing conditions. Therefore, under the coexistence of oxidized sulfur and reduced iron simultaneously, a substantial concentration of soluble arsenic can be present (Korte and Fernando, 1991).

Table 1.1 shows approximate $pK_a$ values for the inorganic arsenic species. It can be concluded that the ionization steps of As(III) and As(V) are significantly different. At normal natural pH environments (pH 4-9), $HAsO_4^{2-}$ and $H_2AsO_4^-$ are the dominant species for As(V) and $H_3AsO_3^0$ is dominant for As(III) (Sadiq, 1997). Based on the data of Table 1.1, the mole fraction of total dissolved As(III) and As(V) as a function of pH can be drawn as shown in Figure 1.3.

**Table 1.1 The $pK_a$ Values of Inorganic Arsenic Species**

| Arsenic species | $pK_1$ | $pK_2$ | $pK_3$ |
|---|---|---|---|
| As(III): Arsenites | 9.2 | 12.1 | - |
| As(V): Arsenates | 2.2 | 6.96 | 11.5 |

Source: Cherry et al, 1979

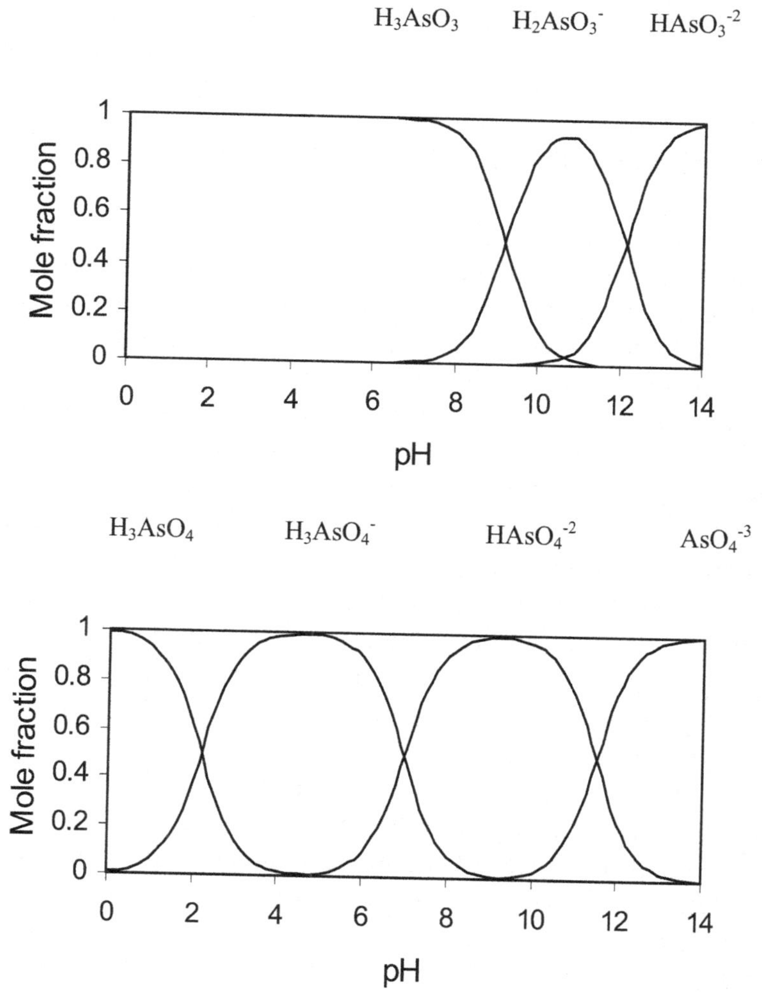


**Figure 1.3. - Mole Fraction of Total Dissolved As as a Function of pH in Water at 25°C: (A) As(III), and (B) As(V)**

## SOURCES OF ARSENIC IN THE ENVIRONMENT

Arsenic occurs naturally and makes up 0.00005 percent of the earth's crust (Pontius et al., 1994), is considered a metalloid (Gulledge and O'Connor, 1973), exhibiting properties of both metals and nonmetals, and is a component of more than 245 minerals. These minerals consist mostly of ores containing sulfide, copper, nickel, cobalt, and lead.

Arsenic enters water by natural means such as volcanic activity, weathering of rocks, and erosion of soils. Agricultural uses of chemicals containing arsenic include pesticides, herbicides and defoliants. Industrial applications of arsenic compounds include wood preservative treatments and decolorizers in glass manufacturing (Gulledge and O'Connor, 1973). Mining, coal burning, copper and lead smelting, and municipal incinerators also release arsenic into the environment. All of these applications provide either a natural or synthetic pathway for arsenic to enter surface water or ground water supplies, thus resulting in contamination.

In natural waters, the most common valence states are arsenate [As(V)] and arsenite [As(III)]. In aerobic surface waters, arsenate is generally more prevalent than arsenite, however arsenite is normally present as well. Higher concentrations of arsenite are more commonly found in anaerobic ground waters (USEPA, 1998). Under certain conditions in the presence of sulfur, arsenate and arsenite no longer comprise the dominant species of arsenic and realgar (AsS) and orpiment ($As_2S_3$) may occur as stable solids (Gupta and Chen, 1978).

## HEALTH ISSUES RELATED TO ARSENIC EXPOSURE

A major pathway of arsenic exposure for United States citizens, resulting from the aforementioned sources, is through consumption and contact of water and food. Another source of exposure to arsenic is via inhalation. Both acute and chronic arsenic exposure, regardless of the pathway, has been found to cause various mild to fatal health effects. The severity of the effects is primarily dependent upon the level of exposure.

Generally, arsenic concentrations are highest in groundwater, especially where geochemical conditions favor arsenic dissolution. However, concentrations of arsenic in surface waters are now of concern. The US Food and Drug Administration has estimated that adults ingest an average of 53 μg/day of arsenic from food. Most of this amount is attributed to seafood, such as marine crabs, lobster, shrimp and cod. The remainder of arsenic intake due to food ingestion is from freshwater fish and shellfish, meats, grains and vegetables. Also, humans are exposed to arsenic via the air, but the average national exposure in the United States has been estimated at 0.006 μg of arsenic per cubic meter, which is considered minimal. However, in areas near smelters or power plants, or other polluted areas, the level of exposure to arsenic due to inhalation may be greater (USEPA, 1998).

In general, inorganic species of arsenic, primarily found in water, are considered more toxic and bio-available than the organic species of arsenic. Long-term chronic exposure to inorganic arsenic can result in cancer of the bladder, lungs, skin, kidneys, nasal passages, liver and prostate. Short-term acute exposure can cause serious circulatory system, nervous system, gastrointestinal complications, and other ill effects. The lethal dose of arsenic to humans is estimated at 1 to 4 mg of arsenic per kilogram for an adult (USEPA, 1998).

Exposure to organic species of arsenic results from ingestion of food and from the metabolism of ingested inorganic arsenic species (USEPA, 1998). Once ingested, soluble forms of arsenic are quickly and readily absorbed by the gastrointestinal tract. The rate of absorption ranges from forty percent to one hundred percent for humans. Once absorbed, arsenic is transported by the blood to various organs in the body. If after ingestion, arsenic is not immediately excreted or absorbed, it will be detoxified by the methylation process, which primarily occurs in the liver (Pontius et al., 1994).

## ARSENIC REGULATIONS AND RESULTING COSTS

In 1975, the USEPA established the MCL for arsenic at 50 ppb, or 0.05 mg/L. This MCL was based on a Public Health Service standard originally established in 1942 (USEPA, 1998). Under the 1996 Amendments to the Safe Drinking Water Act, the USEPA is revising the standard for arsenic and establishing a new MCL.

In 1993, the World Health Organization (WHO) set 10 ppb as the recommended limit for arsenic in drinking water. The European Union, consisting of fifteen nations, adopted this MCL of 10 ppb as a mandatory standard in 1998. Even at this level, it is estimated that lifetime, or chronic, exposure will result in skin cancer for six of every ten thousand people (World Health Organization, 1998).

The USEPA studied the dangers of arsenic exposure for ten years before proposing the 10 ppb standard in June 2000, which is accepted and practiced in many other countries, as previously stated (USEPA, 2001). In the United States, it was estimated that lowering the allowable arsenic level from 50 ppb to 10 ppb would prevent 1000 cases of bladder cancers, as well as 2000 to 5000 lung cancer cases (USEPA, 2000).

The USEPA estimated that the total cost of reducing arsenic contamination to 10 ppb nationwide could be approximately $181 million each year. The average increase in household cost for water that meets the new arsenic standards depends on the size of the water system and the number of people served by that system. In many locations, particularly in the southwestern United States, arsenic contamination is prevalent. USEPA estimates that five percent, or 1,100 water systems in the United States, which serve approximately 2 million people, will have to take measures to meet the standard of 10 ppb. Of all of these affected systems, 97 percent are small systems that serve fewer than 10,000 people.

## METHODS OF ARSENIC REMOVAL FROM DRINKING WATER

Numerous treatment technologies have been developed and applied to arsenic removal from various contaminated waters. Removal of arsenic can be grouped into four categories of processes: precipitative processes such as coagulation/filtration, membrane processes such as reverse osmosis; adsorption processes such as activated alumina and solid phase iron-oxy-hydroxides; and ion exchange (USEPA, 1999). These processes are summarized in Table 1.2.

**Table 1.2 - The Commonly Applied Technologies for Arsenic Removal From Water**

| Technology | Brief Description |
|---|---|
| Precipitative Processes | Coagulation, enhanced coagulation, and lime softening are included this category. Metal hydroxides such as ferric salts, alum, and lime are used as precipitants. Pre-oxidation of As(III) to As(V) might be needed. Affecting factors on performance are precipitant type and dosage, arsenic oxidation state, pH, and competing compounds. |
| Membrane Processes | Nano-filtration (NF) and reverse osmosis (RO) membranes are the only membranes applied without pretreatment due to the low molecular weight of the arsenic species. Generation of a large volume of residuals is expected. The presence of solids and colloids, oxidation state of arsenic, pH, and temperature can be the factors affecting the removal performance. Precipitative processes ahead of micro-filtration and ultra-filtration have also been successfully applied. |
| Adsorptive Processes | The typical adsorbents are activated alumina (AA), activated carbon (AC), and iron-based adsorbents such as, granular ferric hydroxide (GFH), granular ferric oxide (GFO), and iron oxide-coated sand (IOCS). Each adsorbent has different aqueous conditions for optimal arsenic removal. Currently, the most effective adsorbent is GFH. Factors such as pH, oxidation state of the arsenic, competing ions, and empty bed contact time (EBCT) influence the removal efficiency. |
| Ion Exchange | Strong base resins have been typically used. Important factors affecting performance are pH, competing ions, resin type, alkalinity, and arsenic oxidation state. |

**Precipitative Processes**

Due to the traditional use of coagulation in water treatment, the utility of coagulants such as aluminum sulfate, ferric sulfate, and ferric chloride to remove arsenic have been rigorously examined (McNeill and Edwards, 1997; Gulledge and O'Conner, 1973; Scott et al., 1995; Hering et al., 1996; Banerjee et al., 1999). Dissolved inorganics, such as arsenic, can be removed through coagulation by adsorption (association of the dissolved contaminant with the surface of the precipitate), occlusion (entrapment of adsorbed contaminant in the interior of the growing particle), and solid-solution formation (incorporation of the contaminant into the bulk solid phase rather than only onto the surface of the precipitate). However, as the coagulant dosage increases, adsorption onto the metal oxy-hydroxide floc becomes the major mechanism for removal (Hering et al., 1996).

The removal of arsenic is dependant on the speciation, coagulant type, pH, and the presence of competing ions. When ferric chloride was applied as the coagulant, complete removal of As(V) was seen with 5 mg/L of ferric chloride, while complete removal of As(III) was not achieved with as much as 25 mg/L of ferric chloride. Therefore, when As(III) is present, the oxidation of As(III) to As(V) is advantageous in order to effectively remove total arsenic (Hering et al., 1996).

The choice of coagulant is equally important in total arsenic removal. The agreement of many previous researchers was that iron salts, such as ferric chloride and ferric sulfate, were much more effective than aluminum sulfate for the removal of arsenic (Gulledge and O'Conner, 1973; Hering et al., 1996; Scott et al., 1995). Scott et al. (1995) highlighted the study performed by the Metropolitan Water District of Southern California that found over 90% of arsenic present in surface water was removed by 6.5 mg/L ferric chloride, while removal reached only 69% with 20 mg/L of aluminum sulfate, over double the equivalent molar concentration. Research by McNeill and Edwards (1997) attempted to explain the reasoning behind ferric chloride's increased removal over aluminum sulfate. McNeill and Edwards concluded that ferric hydroxide and aluminum hydroxide are equally effective for removal of arsenic. The difference in removal is not due to the better removal of ferric verses aluminum hydroxide floc, but due to the formation of the metal hydroxides from the coagulants and the bonds produced. Ferric chloride almost completely reacts to form ferric oxy-hydroxide, while the presence of free $Al^{3+}$ ions in solution after coagulation with aluminum sulfate tends to support the claim that aluminum sulfate does not react fully. Furthermore, since ferric oxy-hydroxides form stronger bonds with arsenic than aluminum hydroxide, when filtered, arsenic bonded with aluminum hydroxide is more likely to be separated and returned to the solution. Other factors affecting removal are the solution pH as well as the presence of competing ions. In relatively neutral pH ranges (4-9), As(III) is only slightly affected by pH changes with maximum removal between pH 6 and 7. However As(V) removal, after remaining constant between pH 4 and 8, decreases sharply at pH greater than 8. Competing ions such as sulfate also affect arsenic removal. As(III) removal is more effected by sulfate presence than As(V). Models used by Hering et al. (1996) seem to over predict the affects of sulfate on removal, thus requiring more research to verify the correlation between sulfate and coagulation removal of arsenic.

***Iron/Manganese Oxidation***

Similar to coagulation, the oxidation of Fe(II) and Mn(II) leads to the formation of metal hydroxides that can remove soluble arsenic. The production is analogous to "in situ coagulant addition" (Edwards, 1994). In a solution containing an initial arsenic concentration of 10 μg/L, the addition of 2 mg/L of Fe(II) resulted in a final arsenic concentration of 0.75 μg/L. When a smaller amount of Fe(II) was added (1 mg/L), an initial As(V) solution containing 22 μg/L was reduced by 83 percent to about 3.5 μg/L. The treatment of Mn(II) was less effective with the addition of 1 mg/L of Mn(II) resulting in a reduction of only 25% of an initial 25 μg/L As(V) solution. At even higher Mn(II) addition, 3 mg/L reduced a 12 μg/L As(V) solution to 3.75 μg/L (69% removal). Furthermore, if equal amounts of Fe(II) and Mn(II) were added to the same solution, Edwards (1994) determined that removal of arsenic by Fe(II) was much greater, accounting for 92% of the total arsenic removed.

### *Enhanced Coagulation*

In work conducted by Cheng et al. (1994), bench, pilot, and demonstration scale tests were conducted on the use of enhanced coagulation. The conditions studied were to increase the coagulant dosage of aluminum sulfate and ferric chloride from 10 to 30 mg/L and/or decrease the pH from 7 to 5.5. Over 90% removal of As(V) was achieved under enhanced coagulation conditions. With ferric chloride, 96% removal of 5 μg/L arsenic was achieved with 5 mg/L addition, while 90% removal was not achieved when alum was added without lowering the pH. pH adjustment did not affect the removal percentages when ferric chloride was used.

### *Lime Softening*

Hardness is caused by the presence of calcium and magnesium in solution. By adding lime, the carbonate equilibrium is shifted causing both calcium and magnesium to precipitate. Calcium precipitates as calcium carbonate at pH values greater than 9-9.5 while magnesium precipitates as magnesium hydroxide at pH greater than 10.5 (USEPA, 2000a). McNeill and Edwards (1997), as well as others, have shown precipitated calcium and magnesium can be effective in removing arsenic.

Between pH 10.5 and 12, arsenic removal from precipitating calcium carbonate was greater than 40%. Magnesium hydroxide precipitation was more effective, removing up to 100% of the arsenic at pH 11-11.5 (McNeill and Edwards, 1997). In facilities precipitating just calcite, removal was between 0 – 10%, while for facilities employing both calcite and $Mg(OH)_2$ precipitation removal was up to 95% (McNeill and Edwards, 1995). A possible explanation for the more effective removal by magnesium is the magnesium hydroxide's high positive surface charge at pH 11, which attracts the negatively charged arsenate ions. The presence of phosphate as well as carbonate hinders the removal of arsenic by magnesium hydroxide (McNeill and Edwards, 1997).

## **Membrane Processes**

Membrane processes are defined by the size class and are broken into micro-filtration (MF), ultra-filtration (UF), nano-filtration (NF), and reverse osmosis (RO), with the membrane pore size decreasing with each respective size class. Membrane processes can remove arsenic through filtration, electrostatic repulsion, and sorption. While the membrane pore size is the most important variable in determining removal, the chemical characteristics of both the membrane material and arsenic are also important. NF, and RO have both been shown to be effective at removal of dissolved arsenic, while MF and UF, though effective at removal of particulate arsenic, are not viable when used without pretreatment to remove dissolved arsenic (USEPA, 1999).

### *Micro and Ultra-filtration*

Although MF and UF without pretreatment are not effective for removing dissolved arsenic, the combination of UF with pretreatment such as coagulation/co-precipitation with ferric chloride or aluminum sulfate has been shown to be effective (Amy et al., 2002). In the work of

Amy et al., a variety of arsenic contaminated waters were tested and substantial removal documented.

### *Nano-filtration*

NF membranes are commonly used to remove divalent ions ($Ca^{2+}$ and $Mg^{2+}$), but not monovalent salts ($Na^{+}$ and $Cl^{-}$). NF is effective for removing As(V) and As(III) though As(V) is more effectively removed. At pH 7, 12% of As(III) was removed while over 85% of the As(V) present was removed. Therefore, conversion of As(III) to As(V) is advantageous. However, due to its smaller pore size, NF is more prone to fouling than the larger pore size membranes (micro-filtration and ultra-filtration) (Hering and Elimelech, 1996).

### *Reverse Osmosis*

Reverse osmosis (RO) has been traditionally used in the desalination of brackish water and sea water (USEPA, 2000a). RO systems can consistently remove up to 90% of As(V), with total removal possible, and between 60-70% of As(III) (Fox, 1989; Wayna, 1997). A major advantage of RO systems is the overall improvement of the water quality since RO filtration is not specific to arsenic, thus removing other contaminants such as dissolved solids. A disadvantage to an RO system is that it must be operated at high pressures causing increased pumping cost. Also, not only do RO systems produce small quantities of treated water, but retentate rejection is high (Fox, 1989).

## Adsorptive Processes

Alumina ($Al_2O_3$) has been examined extensively for the removal of arsenic, and research indicates that activated alumina is indeed effective for arsenic removal (Fox, 1989). Testing conducted on the arsenic sorption capabilities of activated alumina range from simple batch tests and column studies to pilot test programs at treatment facilities (Gutpa and Chen, 1978; Frank and Clifford, 1986; Rubel and Hathaway 1985). The two species of arsenic, As(V) and As(III), demonstrate vastly different sorption kinetics and equilibrium capacities. Rate studies indicate that As(V) sorbs at a faster rate than As(III). Ghosh and Yuan (1987) observed that the alumina sorbed As(V) much more rapidly than As(III). The maximum removal of As(V) was achieved in less than twenty-four hours at values of pH greater than 6.5. However, pH poses a significant effect on the sorption of arsenic on alumina.

Gupta and Chen (1978) observed that the optimum pH ranged from 4 to 7 for effective sorption of As(V) onto alumina. The predominate species of As(V) between the pH range of 3 to 6.5 is $H_2AsO_4^-$. In the work of Rosenblum and Clifford (1984), the optimum sorption occurred at a pH of 6, where the negatively charged arsenic ions sorb to the positively charged surface of the alumina particles, thus removing the arsenic ions from solution. Since As(V) sorbed in basic solutions, as well as acidic solutions, Ghosh and Yuan (1987) concluded that both electrostatic and specific chemical interactions are associated with the sorption of As(V).

In the work of Gupta and Chen (1987), it was demonstrated that within the pH range of 4 to 9, only minimal sorption of As(III) occurred. Kuhlmeier and Sherwood (1996) concluded that As(III) sorbs twenty times less efficiently than As(V). Between the pH range of 1 to 9, the predominate species of As(III) is $H_3AsO_3^{\circ}$. This form of As(III) has a neutral charge, which explains its reluctance to sorb to the positive surface of the alumina particles. However, above pH 9, $H_2AsO_3^-$ is the primary species of As(III) that is present. The alumina particles begin to develop a negative charge at a pH of 8, and at pH values greater than 9, the surface of the alumina becomes strongly negative, repelling the negatively charged As(III) ion. Because of this effect of pH on As(III) sorption, the As(III) is usually oxidized to As(V) prior to adsorption.

Iron oxide coated sand (IOCS) has also been shown to be an effective media for removal of arsenic (Benjamin et al., 1998). IOCS consists of sand grains coated with ferric hydroxide, which is most often used in fixed bed applications with removal being most effective at low pH values. Increasing the pH from 5.5 to 8.5 decreases removal of As(V) by 30% (Benjamin et al., 1998). The speciation and co-competing ions are also important. As with other media, As(V) is more effectively removed than As(III). In tests performed by Benjamin et al. (1998) for 2 and 24 hours, As(V) adsorption removals were close to 60% while As(III) adsorption removals were only 50%. A major drawback of iron oxide coated sand had been the ineffectiveness of regeneration and recovery of the adsorbed arsenic (Benjamin et al., 1996; Benjamin et al., 1998).

Recently several proprietary iron based sorbents have appeared on the market. These sorbents are believed to be amorphous iron hydroxides and oxides often given the acronym GFH for granular ferric hydroxide and GFO for granular ferric oxide. GFH and GFO are used in treatment works in a manner similar to alumina, typically in packed beds and columns. Residuals examined in this study include both spent GFH and GFO.

### Ion Exchange

Ion exchange has long been recognized as an effective treatment process for the removal of arsenic. Strong-base anion exchange resins will remove As(V) which is present in near neutral pH ranges as an anion. As(III) is poorly removed by ion exchange due to its neutral charge in pH ranges of natural water, and thus pre-oxidation of As(III) to As(V) is required for removal (Chwirka et al., 2000; Fox, 1989). Furthermore, since strong-base ion exchange resins often are not selective in their removal, the presence of competing ions such as nitrate and sulfate hinder removal efficiency. According to Frey et al. (1998), if sulfate concentrations are greater than 120 mg/L, removal of arsenic by ion exchange is not economically feasible due to the need to change and/or regenerate the resin bed frequently.

## CHARACTERIZATION OF ARSENIC TREATMENT RESIDUALS

Limited work has been conducted on the characterization and stabilization of arsenic containing residuals. Cornwell et al. (2003) surveyed state and federal regulations and summarized data on composition of residuals and their performance in leaching tests. They also evaluated some stabilization methods and developed a decision tree for management of arsenic-containing residuals. Amy et al. (2000) investigated arsenic leachability in various water treatment plant (WTP) residuals by the toxicity characteristic leaching procedure (TCLP). The results showed that arsenic concentrations in the residuals generated from WTPs varied greatly as did the soluble arsenic concentrations in the leachates. However, the concentrations of arsenic in the leachates were not over the limit of 100 times of MCL for arsenic. Thus, the landfill disposal alternative was recommended for arsenic containing WTP residuals.

## TOXICITY CHARACTERISTICS LEACHING PROCEDURE (TCLP)

The TCLP is a regulatory test that is employed to determine whether a particular contaminant can be disposed in a landfill (USEPA, 1992). Evaluation of various stabilization techniques has been broadly conducted using the TCLP method (LaGrega et al., 1994). In the TCLP method, target solid materials are crushed to small size, mixing with an extraction liquid, and agitated in a rotary reactor for 18 hours. Subsequently, insoluble and soluble phases are separated through a filter (USEPA, 1992). However, it was reported that TCLP would not be a suitable leaching test to predict the release of contaminants (oxo-anion-forming elements) from municipal solid wastes (Hooper et al., 1998). Loeppert et al. (2003) suggested that arsenic leaching could be higher than that predicted by the TCLP test. The TCLP is widely used for regulatory purposes, but other leaching test procedures exist and provide different levels of arsenic release. For example, the Ca WET procedure has been reported to extract much higher levels of arsenic from residuals from water treatment process residuals because of its longer extraction period, different leaching solution (citric acid buffer) and anaerobic conditions (Cornwell et al., 2003).

## SOLIDIFICATION AND STABILIZATION (S/S)

USEPA (2002) introduced several treatment technologies for arsenic in soils and wastes: solidification and stabilization (S/S), vitrification, soil washing/acid extraction, pyro-metallurgical recovery, and in situ soil flushing. S/S has been widely applied to arsenic treatment (immobilization) in soils (Miller, 1996; Voigt et al., 1996), industrial wastes (Fuessle and Taylor, 2000; Palfy et al., 1999; Yaziz et al., 1999; Dutre and Vandecasteele, 1998; Dutre and Vandecasteele, 1995), and residuals (Kameswari et al., 2001; Roy et al., 1992). In practice, only solidification and stabilization can be applicable to arsenic treatment in WTP residuals. Arsenic behavior in soils and wastes are similar to residuals since both are solid-phase media, thus S/S processes can be utilized for arsenic immobilization in residuals. Through physical and chemical means by S/S, hazardous substances and contaminants are stabilized within hardened structures. In general, S/S is designed to achieve one or more of the following: (a) improve the handling and physical characteristics of wastes, (b) decrease the surface area of the waste mass, (c) limit the solubility of hazardous contaminants, and (d) detoxify contained contaminants (LaGrega et al., 1994).

Several additives (binders) may be employed in the stabilization processes. Commonly used additives are cement (typically ordinary Portland cement (OPC)), lime, pozzolans such as, fly ash, slag, and kiln dust, and organically modified clays (LaGrega et al., 1994). The following materials have frequently been used as binders and reagents for the stabilization of arsenic tainted soils and wastes: Portland cement, lime, ferrous sulfate, ferric salts, and fly ash.

Portland cement is the most frequently used binder for solidification/stabilization (USEPA, 2002). Type I OPC typically consists of about 50% of tricalcium silicate ($3CaO{\bullet}SiO_2$), about 25% of dicalcium silicate ($2CaO{\bullet}SiO_2$), about 10% of tricalcium aluminate ($3CaO{\bullet}Al_2O_3$), about 10% of tetracalcium aluminoferrite ($4CaO{\bullet}Al_2O_3Fe_2O_3$), and 5% other oxides by a mass basis (Conner, 1993). Tricalcium silicate and dicalcium silicate hydration can be expressed by the following reactions (Taylor, 1997):

$$2(3CaO \bullet SiO_2) + 6H_2O \rightarrow 3CaO \bullet 2SiO_2 \bullet 3H_2O + 3Ca(OH)_2 \quad (1.7)$$

$$2(2CaO \bullet SiO_2) + 4H_2O \rightarrow 3CaO \bullet 2SiO_2 \bullet 3H_2O + 3Ca(OH)_2 \quad (1.8)$$

It takes approximately 1 year to achieve 95 to 98% cement hydration. Calcium silicate hydrate (C-S-H) and portlandite ($Ca(OH)_2$) are the two main products of cement hydration. C-S-H represents about 60 to 70%, $Ca(OH)_2$ about 20 to 25%, and the other solid phases account for about 5 to 15% of the cement hydrates (Glasser, 1993).

Lime (calcium oxide, CaO) has been used as a material to stabilize arsenic contaminated wastes, residuals, and soils (Kim et al., 2003; Kameswari et al., 2001; Bothe and Brown, 1999; Dutre and Vandecasteele, 1998). The calcium arsenate precipitates ($Ca_4(OH)_2(AsO_4)_2 \bullet 4H_2O$, $Ca_5(AsO_4)_3OH$, and $Ca_3(AsO_4)_2 \bullet 2H_2O$) are generated as stable compounds for As(V) immobilization (Bothe and Brown, 1999). Dutre and Vandecasteele (1998) added cement and calcium to minimize the arsenic leaching in industrial wastes containing arsenic in high concentrations. They explained that the main reason why arsenic concentration in the leachate was lowered enormously was that $CaHAsO_3$, a stable and less soluble calcium-arsenic compound, was formed. Kim et al. (2003) studied the stabilization of arsenic in mine tailings using iron. The results indicated that lower arsenic concentration in the leachate at a pH range of 3 to 6 was observed when $Ca(OH)_2$ rather than NaOH was used for pH adjustment. When higher initial Fe(III) concentration was added, this trend was more obvious. However, the solidified solids treated by lime only will decompose slowly when exposed to atmospheric $CO_2$ to form $CaCO_3$ and soluble arsenic species (Riveros et al., 2001).

Ferrous sulfate ($FeSO_4 \bullet 4H_2O$) addition to arsenic contaminated solids produces ferric arsenate ($FeAsO_4$), an insoluble compound or ferric hydroxide ($Fe(OH)_3$) precipitates which adsorb arsenic. However, the ferric arsenate generation mechanism by ferrous iron has not been well documented (Voigt et al., 1996). Sandesara (1978) recommended the use of ferrous sulfate rather than ferric sulfate due to the interferences with the curing of the cement. Indeed, ferrous sulfate has been employed for large-scale remediation of arsenic contaminated sites (USEPA, 1998; Miller, 1996). Sulfate can adversely influence the solids characteristics of wastes/cement matrices. Sulfate ions can result in ettringite formation, which causes the development of cracks in solidified solids (Taylor, 1997). Other ferric salts were examined by Kim et al. (2003) to stabilize arsenic in mine tailings and concluded that the effectiveness of ferric sulfate is better than that of ferrous sulfate. However, ferrous S/S was recommended due to the greater solubility of ferrous hydroxide and the generation of larger molecules of ferric hydroxy-arsenic complexes than ferrous arsenic compounds (Fuessle and Taylor, 2000).

Fly ash is an additive that has been employed with other binders such as cement and lime (Kameswari et al., 2001; Fuessle and Taylor, 2000). Akhter et al. (1997) tested solidification/stabilization of arsenic using various sets of binders over a period of four years. The performance of the combination of OPC and fly ash (Class F) was significantly lower and the leached arsenic concentration of the OPC and fly ash matrix became progressively greater over time.

In summary, a variety of low cost, common additives have been examined in some form for the immobilization of arsenic. Of these compounds OPC and lime appear to be the most successful. However, iron salts and fly ash also show some promise. The work of this report will elucidate the differences between these solidification and stabilization compounds; a primary objective.

# CHAPTER 2: SORPTION AND DESORPTION STUDIES USING RESIDUALS OF KNOWN COMPOSITION

## INTRODUCTION

The first phase of this research concerns the removal of arsenic using coagulating sorptive chemicals of known composition. This phase was conducted in order to obtain representative samples of treatment residuals contaminated with arsenic. The influence of colloidal material in the residuals and their influence on arsenic leaching was an additional objective. Two coagulants were the main focus of this portion of the study; aluminum sulfate or alum ($Al_2(SO_4)_3$) and ferric chloride ($FeCl_3$). Their selection was based on the wide use of these compounds in the water treatment industry, their known ability to remove arsenic, and the fact that the oxy-hydroxides produced by either iron or aluminum are also used as granular sorbents. Thus, the sorption and desorption seen using ferric chloride or alum should be similar to that of granular iron sorbents and alumina. Additionally, calcium carbonate was examined for both sorption and desorption behavior.

Synthetic arsenic residuals were obtained by adding ferric chloride or alum at a 20:1 mole ratio to a 100 mg/L solution of either arsenate or arsenite. Additionally, the influence of colloidal material on arsenic sorption and desorption was examined. The colloidal additives included kaolinite (a 1:1 clay mineral), montmorillonite (a 2:1 clay mineral), and organic material produced by adding 100 mg/L carboxymethylcellulose and 100 mg/L of humic acid. The choice of carboxymethylcellulose and humic acid was made in order to provide both a soluble and insoluble organic component. Calcium carbonate was also used as an arsenic sorbent, however only half of the arsenic concentration used in the other formulations was used. These reaction mixtures include:

1) Ferric chloride and arsenate;
2) Ferric chloride, arsenate and kaolinite;
3) Ferric chloride, arsenate and montmorillonite;
4) Ferric chloride, arsenate, carboxymethylcellulose and humic acid;
5) Ferric chloride and arsenite;
6) Ferric chloride, arsenite and kaolinite;
7) Ferric chloride, arsenite and montmorillonite;
8) Ferric chloride, arsenite, carboxymethylcellulose and humic acid;
9) Alum and arsenate;
10) Alum, arsenate and kaolinite;
11) Alum, arsenate and montmorillonite;
12) Alum, arsenate, carboxymethylcellulose and humic acid;
12) Alum and arsenite;
13) Alum, arsenite and kaolinite;
14) Alum, arsenite and montmorillonite;
15) Alum, arsenite, carboxymethylcellulose and humic acid;
16) Calcium carbonate and arsenate;
17) Calcium carbonate and arsenite.

## EXPERIMENTAL METHODS AND PROCEDURES

### Preparation of Residuals

Laboratory residuals were prepared by mixing the various compounds to obtain a total of ten liters of solution for each batch. The exact composition of each solution is given below. The pH was continuously monitored and adjusted by adding NaOH or $HNO_3$ to maintain the pH between 7 and 8 for ferric chloride and alum coagulation. After adding the chemicals, the mixtures were agitated for about 30 minutes and subsequently allowed to settle for an additional 30 minutes. After sedimentation a sample from the surface of the vessel was drawn, filtered through a 0.2 μm membrane, and analyzed for arsenic concentration. The remaining solution was decanted and filtered through a Buchner funnel lined with filter paper. The residuals were produced with both As(III) and As(V) according to the following solution chemistry.

#### *Iron sorption*

The solution for these experiments contained an initial arsenic concentration of 100 mg/L and was prepared by diluting 0.2 L of 5 g/L arsenic stock solution to 10 L in DI water and stirring vigorously for 5 minutes. The $FeCl_3$ concentration was 26.70 mM. This was accomplished by adding 0.2 L of 1.335 M $FeCl_3$ while stirring. The pH was monitored and maintained at 7-8 by adding NaOH. After adding all chemicals, the solution was gently stirred for about 30 minutes, then allowed to settle for an additional 30 minutes.

$$\frac{100mg/L}{74.9216g/mol} \times \frac{1g}{1000mg} = 1.33473mM \text{ of arsenic}$$

$$\frac{mole\_of\_arsenic}{mole\_of\_Fe} = \frac{1.33473mM}{26.70mM} = 0.05$$

#### *Aluminum sorption*

The initial arsenic concentration was also 100 mg/L and was prepared by diluting 0.2 L of 5 g/L arsenic stock solution to 10 L in DI water and stirred vigorously for 5 minutes. The $Al_2(SO_4)_3$ concentration was 13.348 mM. This was accomplished by adding 0.2 L of 0.6674 M $Al_2(SO_4)_3$ while stirring. The pH was monitored and maintained at 7-8 by adding NaOH. After adding all chemicals, the solution was gently stirred for about 30 minutes, then allowed to settle for an additional 30 minutes.

$$\frac{mol\_of\_arsenic}{mol\_of\_Al} = \frac{1.33473mM}{13.348 \times 2mM} = 0.05$$

#### *Clay Minerals*

A suspension of clay minerals was prepared by adding 200 mg/L of either kaolinite or montmorillonite to a solution of 100 mg/L arsenic in DI water. The suspension was then coagulated with either alum or ferric chloride using the same procedure given above, and the pH was also adjusted as above.

***Organic Materials***

A solution with 100 mg/L arsenic, 100 mg/L carboxymethylcellulose and 100 mg/L of humic acid was prepared in DI water. This mixture was coagulated with either alum or ferric chloride using the same procedure as above.

***Calcium Carbonate:***

The initial arsenic concentration was 50 mg/L. Then 300 mM of both $Ca(OH)_2$ and $Na_2CO_3$ were added to 9.9 L of DI water. This reaction mixture was stirred and allowed to settle for 30 minutes.

**Determination of Total Arsenic and Fe/Al/Ca Concentration**

Each residual sample was examined for elemental composition using the digestion procedures given in Appendix E. In general these digestion procedures were as follows. Approximately 1 gram of sample was added to the digestion vessel. The exact sample weight was recorded to the nearest 0.001 g. Ten milliliters of 1:1 $HNO_3$ were added to each sample in a fume hood. The sample was heated to 95 ± 5 °C for 10 to 15 min without boiling. The sample was subsequently allowed to cool, and an additional 5 mL of concentrated $HNO_3$ was added and the sample reheated for 10 to 15 minutes. This procedure was repeated until no brown fumes were generated indicating the complete reaction with $HNO_3$. After the sample had cooled, 2 mL of 30% $H_2O_2$ was added to each sample, and the sample was heated until effervescence subsided and then allowed to cool. Additional 30% $H_2O_2$ in 1 mL aliquots was added while warming continuously, until the effervescence was minimal or until the general sample appearance was unchanged. No more than a total of 10 mL 30% $H_2O_2$ was added. After cooling, the samples were diluted to 50 mL with DI water. Particulates in the digestion mixture were removed by filtration. The samples were then analyzed by AAS.

**TCLP Test Procedure**

The residuals produced were stored in acid washed polyethylene bottles under refrigeration until their use. After production the residuals were tested for desorption using a variety of leachant solutions. The first method examined was the TCLP. The exact procedure for conducting the TCLP protocol is given in Appendix B. Basically, residual samples were weighed and dried at 105 °C to determine percent dry solids. The TCLP tests were then conducted by mixing residual samples with the TCLP extraction solutions in capped polyethylene bottles. TCLP extraction fluid #1 was used for this portion of the investigation at a leachant to residuals (wet solid basis) mass ratio of 20: 1. Therefore, for 10 g of wet residuals, 200 mL of leachant were used. TCLP extraction kinetic experiments were also conducted. Aliquots of suspension samples were collected at several time intervals (0.5, 1, 2, 4, 8, 18 hr) and filtered through a 0.2 μm pore size membrane filter. Subsequently, all samples were analyzed for soluble arsenite and arsenate using hydride generation with atomic absorption spectroscopy (HG-AAS).

### Desorption Versus pH

Solution pH plays a significant role in determining the extent of arsenic sorption/desorption from the surface of metal oxy-hydroxides. To explore the influence of pH on arsenic desorption from the synthesized residuals, 10 g of each sample was added to 200 mL of DI water with the pH adjusted by adding either NaOH or $HNO_3$ as needed to achieve the desired final pH. An ionic strength buffer was added to each mixture to give a 10 mM $NaNO_3$ concentration. The range of pH investigated was 4, 6, 8, and 10. Residuals and DI water were mixed in polyethylene bottles by end-over-end rotation, and aliquots of suspension samples were collected at two time periods (8 and 24 hr). The collected samples were filtered through 0.2 μm pore size membranes and the resulting filtrate was analyzed for arsenic concentration by HG-AAS.

### Desorption Versus Competing Ions

The presence of background ions has been shown to adversely influence the sorption of compounds such as arsenic on the surfaces of minerals. Phosphate tends to compete with arsenic on the surface of the oxy-hydroxides for sorption sites and thus, tends to desorb the arsenic compounds. To examine the impact of phosphate on the synthesized residuals, a 1.33 mM $NaH_2PO_4{\bullet}H_2O$ and 10 mM $NaNO_3$ solution was prepared and adjusted to pH 7 using NaOH or $HNO_3$. Sulfate has also been suspected of inhibiting arsenic sorption in coagulation systems. A solution similar to the phosphate mixture was formulated using 1.33 mM $Na_2SO_4$ and 10 mM $NaNO_3$ and adjusting the solution to pH 7 using NaOH or $HNO_3$. The influence of chloride was examined using a 1.33 mM NaCl and 10 mM $NaNO_3$ solution at a pH of 7. Ten grams of each synthetic residual (wet basis) and 200 mL of the phosphate, sulfate or chloride solution were added to acid washed polyethylene bottles and mixed by end over end rotation. Samples were taken at time intervals of 0.5, 1, 2, 4, 8 and 18 hours. All of the desorption studies were conducted in triplicate. The withdrawn samples were filtered using a 0.2 μm membrane filter and analyzed using HG-AAS.

## RESULTS AND DISCUSSION

### Arsenic Removal by Various Coagulants and Background Compounds

The first portion of the investigation concerns the removal of each arsenic compound from laboratory-synthesized solutions. These results appear in Tables 2.1 and 2.2. The oxy-hydroxides formed by the addition of the iron salt ferric chloride are obviously better sorbents than either alum or calcium carbonate at removing arsenite. Alum does a reasonable job of removing arsenate and, to some extent, so does the calcium carbonate formed during lime/soda softening. The improved sorption of the iron oxy-hydroxides is most likely due to the iron oxide product, primarily ferrihydrite, which is a strong arsenic sorbent with a net positive surface charge at low to neutral pH. Aluminum tends to form a large number of oxy-hydroxides, not all of which are strong sorbents due to lower reactive surface areas.

**Table 2.1. Arsenite Removal for Each Mixture Remaining in Solution**

| Coagulant/Chemicals | Soluble As(III) Concentration (µg/L) | Initial As(III) Concentration (µg/L) | Removal Efficiency (%) |
|---|---|---|---|
| FeCl3 | 33 | 100,000 | >99 |
| FeCl3 + Kaolinite | 32 | 100,000 | >99 |
| FeCl3+Montmorillonite | 35 | 100,000 | >99 |
| FeCl3 + Organic Matter | 59 | 100,000 | >99 |
| Alum | 51,000 | 100,000 | 49 |
| Alum + Kaolinite | 56,000 | 100,000 | 44 |
| Alum + Montmorillonite | 63,000 | 100,000 | 37 |
| Alum + Organic Matter | 58,000 | 100,000 | 42 |
| Ca(OH)2 + Na2CO3 | 48,000 | 50,000 | 3.3 |

**Table 2.2. Arsenate Removal for Each Mixture Remaining in Solution**

| Coagulant/Chemicals | Soluble As(III) Concentration (µg/L) | Initial As(III) Concentration (µg/L) | Removal Efficiency (%) |
|---|---|---|---|
| FeCl3 | BDL | 100,000 | >99 |
| FeCl3 + Kaolinite | BDL | 100,000 | >99 |
| FeCl3+Montmorillonite | 4.4 | 100,000 | >99 |
| FeCl3 + Organic Matter | 7.8 | 100,000 | >99 |
| Alum | 22 | 100,000 | >99 |
| Alum + Kaolinite | 18 | 100,000 | >99 |
| Alum + Montmorillonite | 24 | 100,000 | >99 |
| Alum + Organic Matter | 26 | 100,000 | >99 |
| Ca(OH)2 + Na2CO3 | 19,000 | 50,000 | 62 |

The physical composition of each residuals sample is shown in Tables 2.3 and 2.4. It is interesting to note that alum and lime/soda residuals have lower moisture content (higher % dry solids) than the iron based residuals. This trend was especially true for the lime residuals. The iron and aluminum were added in molar amounts, and Fe has a molecular weight of 56 (gm/mole), while that of aluminum is 27 (gm/mole), accounting for the differences shown in Tables 2.3 and 2.4. From a mass balance stand point, the recoveries were well within the values expected. The only major concern is the inability of the calcium system to sorb arsenic. Several research papers have examined lime/soda softening for possible arsenic removal, and these data indicate that calcium carbonate is a poor sorbent for both arsenate and arsenite and that a substantial amount of magnesium must be present in the hard water to adequately remove arsenic. The presence of colloidal additives had little, if any effect, on the sorption of arsenic. The authors initially felt that the presence of the 2:1 clay mineral montmorillonite would improve removal, but these data indicate that no advantage occurs.

**Table 2.3. Physical Properties of Arsenite Residuals**

| Sludge Type | Percent dry solids (%) | Total As (g/kg dry solids) | Total Fe (g/kg dry solids) | Total Al (g/kg dry solids) | Total Ca (g/kg dry solids) |
|---|---|---|---|---|---|
| $FeCl_3$ | 10 | 35 | 510 | - | - |
| $FeCl_3$ + Kaolinite | 9.1 | 32 | 510 | - | - |
| $FeCl_3$+Montmorillonite | 12 | 31 | 490 | - | - |
| $FeCl_3$ + Organic Matter | 9.5 | 30 | 490 | - | - |
| Alum | 16 | 20 | - | 270 | - |
| Alum + Kaolinite | 15 | 16 | - | 240 | - |
| Alum + Montmorillonite | 17 | 15 | - | 230 | - |
| Alum + Organic Matter | 13 | 16 | - | 230 | - |
| $Ca(OH)_2$ + $Na_2CO_3$ | 59 | 0.43 | - | - | 290 |

**Table 2.4. Physical Properties of Arsenate Residuals**

| Sludge Type | Percent dry solids (%) | Total As (g/kg dry solids) | Total Fe (g/kg dry solids) | Total Al (g/kg dry solids) | Total Ca (g/kg dry solids) |
|---|---|---|---|---|---|
| $FeCl_3$ | 11 | 270 | 530 | - | - |
| $FeCl_3$ + Kaolinite | 13 | 250 | 510 | - | - |
| $FeCl_3$+Montmorillonite | 13 | 230 | 490 | - | - |
| $FeCl_3$ + Organic Matter | 10 | 230 | 480 | - | - |
| Alum | 14 | 230 | - | 240 | - |
| Alum + Kaolinite | 15 | 220 | - | 230 | - |
| Alum + Montmorillonite | 18 | 220 | - | 230 | - |
| Alum + Organic Matter | 14 | 200 | - | 210 | - |
| $Ca(OH)_2$ + $Na_2CO_3$ | 60 | 0.79 | - | - | 290 |

**Toxicity Characteristic Leaching Test Extraction and Kinetics**

TCLP tests were conducted on all residuals, and the data is summarized in Tables 2.5 and 2.6. The kinetic data on the leaching of arsenite and arsenate under the TCLP protocol is also displayed graphically in Figures 2.1 and 2.2. These data and data from other leaching experiments are expressed in terms of the mass leached per mass dry solids. It can be converted into the concentration observed in the leaching solution, by dividing by the ratio of liquid volume to dry solid mass (20 L/kg). For example, the amount of arsenite released from the alum residual at 0.5 hr is 100 mg/kg (Table 2.5). This corresponds to a concentration of 5 mg/L. In Figure 2.1, the amount of arsenite desorbed tends to increase to an asymptote over time for both the aluminum and calcium residuals. Arsenate desorption (Figure 2.2) tended to decrease, with the exception of the calcium residuals which exhibited no distinct trend. The ferric chloride residuals showed minimal desorption under the TCLP conditions. Calcium carbonate under the TCLP analysis released a substantial amount of arsenite, and the alum residuals allowed for excessive arsenite leaching. Reaction time appears to be only a factor with the release of arsenite from the calcium and alum based residuals. This time dependent release may be due to the solubility of both aluminum and calcium residuals under the TCLP conditions.

**Table 2.5. Absolute Amounts of Arsenite Released During TCLP at Various Times (unit: mg As/kg dry solids)**

| Sludge Type | Time (hr) | | | | | |
|---|---|---|---|---|---|---|
| | 0.5 | 1 | 2 | 4 | 8 | 18 |
| $FeCl_3$ | 1.3 | 1.3 | 1.2 | 1.1 | 0.97 | 0.94 |
| $FeCl_3$ + Kaol. | 1.3 | 1.2 | 1.1 | 1.1 | 0.97 | 0.87 |
| $FeCl_3$+Mont. | 1.6 | 1.5 | 1.5 | 1.3 | 1.3 | 1.2 |
| $FeCl_3$ + Organic | 1.4 | 1.3 | 1.3 | 1.2 | 1.1 | 1.1 |
| Alum | 100 | 140 | 180 | 220 | 240 | 250 |
| Alum + Kaol | 88 | 130 | 160 | 180 | 190 | 200 |
| Alum + Mont | 92 | 160 | 190 | 210 | 230 | 240 |
| Alum + Organic | 79 | 110 | 140 | 160 | 180 | 190 |
| $Ca(OH)_2 + Na_2CO_3$ | 50 | 56 | 61 | 61 | 63 | 68 |

**Table 2.6. Relative Amounts of Arsenite Released During TCLP at Various Times (unit: %)**

| Sludge Type | Time (hr) | | | | | |
|---|---|---|---|---|---|---|
| | 0.5 | 1 | 2 | 4 | 8 | 18 |
| $FeCl_3$ | <0.1% | <0.1% | <0.1% | <0.1% | <0.1% | <0.1% |
| $FeCl_3$ + Kaol. | <0.1% | <0.1% | <0.1% | <0.1% | <0.1% | <0.1% |
| $FeCl_3$+Mont. | <0.1% | <0.1% | <0.1% | <0.1% | <0.1% | <0.1% |
| $FeCl_3$ + Organic | <0.1% | <0.1% | <0.1% | <0.1% | <0.1% | <0.1% |
| Alum | 0.50% | 0.70% | 0.90% | 1.10% | 1.20% | 1.25% |
| Alum + Kaol | 0.55% | 0.81% | 1.00% | 1.13% | 1.19% | 1.25% |
| Alum + Mont | 0.61% | 1.07% | 1.27% | 1.40% | 1.53% | 1.60% |
| Alum + Organic | 0.49% | 0.69% | 0.88% | 1.00% | 1.13% | 1.19% |
| $Ca(OH)_2 + Na_2CO_3$ | 11.6% | 13.0% | 14.2% | 14.2% | 14.7% | 15.8% |

One interesting trend is apparent in Figures 2.1 and 2.2. In Figure 2.1, the amount of arsenite desorbed tends to increase to an asymptote over time for both the aluminum and calcium residuals. In Figure 2.2 arsenate desorption tends to decrease with time except for the calcium residuals which exhibit no clear trend. This behavior of the arsenate/calcium residuals might be a manifestation of the original quantity of arsenic desorbed (see Table 2.7 and 2.8). These results indicate that the assumption of the aluminum and calcium residual decomposition might be incorrect.

**Table 2.7. Absolute Amounts of Arsenate Released During TCLP at Various Times (unit: mg As/kg dry solids)**

| Sludge Type | Time (hr) | | | | | |
|---|---|---|---|---|---|---|
| | 0.5 | 1 | 2 | 4 | 8 | 18 |
| $FeCl_3$ | 0.011 | 0.011 | 0.008 | BDL | BDL | BDL |
| $FeCl_3$ + Kaol. | 0.012 | 0.012 | 0.009 | 0.007 | BDL | BDL |
| $FeCl_3$+Mont. | 0.017 | 0.017 | 0.016 | 0.011 | 0.009 | BDL |
| $FeCl_3$ + Organic | 0.018 | 0.014 | 0.011 | 0.009 | 0.013 | 0.007 |
| Alum | 0.042 | 0.047 | 0.028 | 0.046 | 0.03 | 0.026 |
| Alum + Kaol | 0.053 | 0.036 | 0.032 | 0.032 | 0.026 | 0.024 |
| Alum + Mont | 0.052 | 0.087 | 0.034 | 0.078 | 0.039 | 0.028 |
| Alum + Organic | 0.03 | 0.03 | 0.025 | 0.027 | 0.027 | 0.023 |
| $Ca(OH)_2 + Na_2CO_3$ | BDL | 0.16 | BDL | 0.058 | 0.28 | 0.072 |

*BDL – Below Detection Limit

**Table 2.8 Relative Amounts of Arsenate Released During TCLP at Various Times (unit: %)**

| Sludge Type | Time (hr) | | | | | |
|---|---|---|---|---|---|---|
| | 0.5 | 1 | 2 | 4 | 8 | 18 |
| $FeCl_3$ | <0.1% | <0.1% | <0.1% | BDL | BDL | BDL |
| $FeCl_3$ + Kaol. | <0.1% | <0.1% | <0.1% | <0.1% | BDL | BDL |
| $FeCl_3$+Mont. | <0.1% | <0.1% | <0.1% | <0.1% | <0.1% | BDL |
| $FeCl_3$ + Organic | <0.1% | <0.1% | <0.1% | <0.1% | <0.1% | <0.1% |
| Alum | <0.1% | <0.1% | <0.1% | <0.1% | <0.1% | <0.1% |
| Alum + Kaol | <0.1% | <0.1% | <0.1% | <0.1% | <0.1% | <0.1% |
| Alum + Mont | <0.1% | <0.1% | <0.1% | <0.1% | <0.1% | <0.1% |
| Alum + Organic | <0.1% | <0.1% | <0.1% | <0.1% | <0.1% | <0.1% |
| $Ca(OH)_2 + Na_2CO_3$ | BDL | <0.1% | BDL | <0.1% | <0.1% | <0.1% |

*BDL – Below Detection Limit

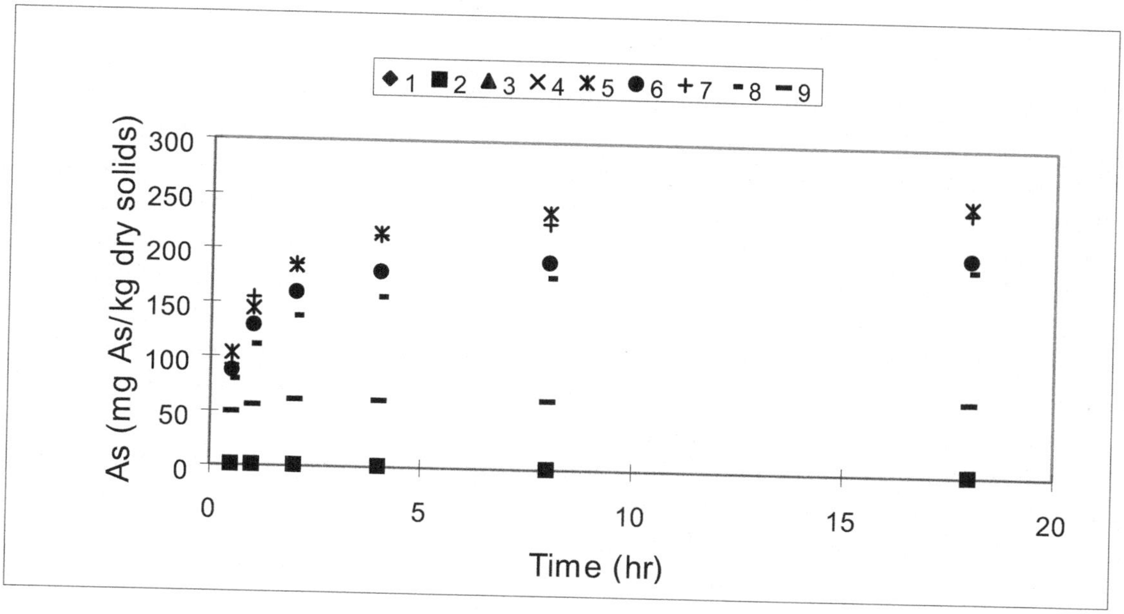


1-$FeCl_3$, 2- $FeCl_3$+Kaol, 3- $FeCl_3$+Mont, 4- $FeCl_3$+Organic, 5-Alum, 6-Alum+Kaol, 7-Alum+Mont, 8-Alum+Organic, 9- $Ca(OH)_2$+$Na_2CO_3$

**Figure 2.1. TCLP Extraction Kinetics for Arsenite Containing Residuals**

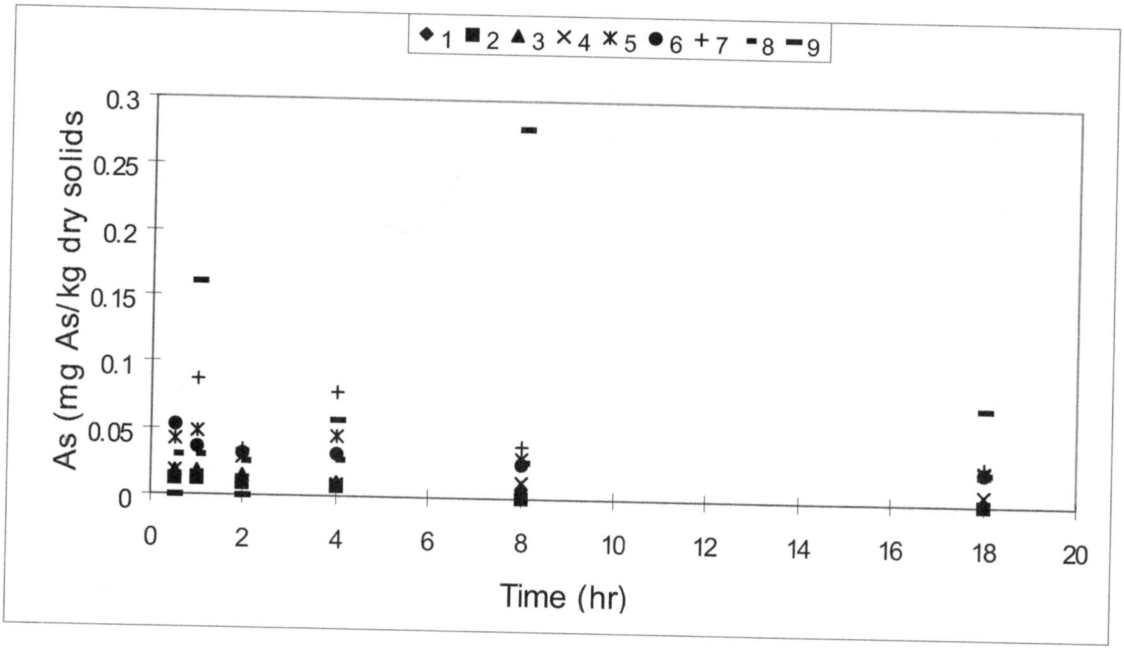


1-$FeCl_3$, 2- $FeCl_3$+Kaol, 3- $FeCl_3$+Mont, 4- $FeCl_3$+Organic, 5-Alum, 6-Alum+Kaol, 7-Alum+Mont, 8-Alum+Organic, 9- $Ca(OH)_2$+$Na_2CO_3$

**Figure 2.2. TCLP Extraction Kinetics for Arsenate Containing Residuals**

**Arsenic Desorption Versus pH**

Tables 2.9 to 2.12 display desorption data for arsenite and arsenate at varying pH. Time dependency is also shown in all tables. The behavior of the iron-based residuals is exactly as expected, where arsenate is shown to desorb at elevated pH while arsenite desorbs at low pH. The interesting item with the iron residuals is how well they resist leaching of arsenic at all pH conditions. In contrast, aluminum residuals released a substantial amount of arsenite. This could be due to arsenite having a weaker affinity for the surfaces of the aluminum solid phases. This is supported by the lower amounts of arsenite that were removed from solution during preparation of the residuals (Table 2.1) The extent to which arsenite was released during desorption experiments was so high that the experiments were repeated three times, but the same result was obtained each time. As arsenite is neutrally charged over most of the pH range that was tested, perhaps that would explain the high levels of desorption. However, even at a pH of 10, where arsenite is predominately negative in charge, extreme desorption is seen. This result implies that the surface of the aluminum oxy-hydroxides must also be negatively charged at pH 10. Thus, some investigation into the speciation and surface chemistry of the aluminum oxy-hydroxides is required.

**Table 2.9. Absolute Amounts of Arsenite Released at Various pH and Leaching Times (units: mg As/kg dry solids)**

| Sample | pH 4 | | pH 6 | | pH 8 | | pH 10 | |
|---|---|---|---|---|---|---|---|---|
| | 8 hr | 24 hr | 8 hr | 24 hr | 8 hr | 24 hr | 8 hr | 24 hr |
| $FeCl_3$ | 25 | 26 | 13 | 15 | 7.7 | 8.6 | 9.4 | 9.7 |
| $FeCl_3$+Kaol | 28 | 36 | 15 | 16 | 8.4 | 9.5 | 10 | 11 |
| $FeCl_3$+Mont | 27 | 33 | 16 | 16 | 9.9 | 10 | 8.2 | 9.7 |
| $FeCl_3$+Org | 27 | 37 | 18 | 19 | 12 | 13 | 12 | 12 |
| Alum | 11,000 | 12,000 | 6,900 | 7,300 | 4,400 | 4,100 | 2,800 | 4,400 |
| Alum+Kaol | 6,600 | 8,500 | 5,400 | 6,200 | 2,900 | 2,500 | 2,500 | 3,500 |
| Alum+Mont | 6,500 | 7,300 | 4,400 | 5,100 | 2,600 | 2,600 | 3,000 | 3,300 |
| Alum+Org | 12,000 | 12,000 | 7,900 | 8,200 | 5,000 | 5,000 | 3,200 | 4,300 |
| $Ca(OH)_2+Na_2CO_3$ | 300 | 310 | 260 | 260 | 160 | 180 | 100 | 110 |

**Table 2.10. Relative Amounts of Arsenite Released at Various pH and Leaching Times (unit: %)**

| Sample | pH 4 | | pH 6 | | pH 8 | | pH 10 | |
|---|---|---|---|---|---|---|---|---|
| | 8 hr | 24 hr | 8 hr | 24 hr | 8 hr | 24 hr | 8 hr | 24 hr |
| $FeCl_3$ | <0.1% | <0.1% | <0.1% | <0.1% | <0.1% | <0.1% | <0.1% | <0.1% |
| $FeCl_3$+Kaol | <0.1% | 0.11% | <0.1% | <0.1% | <0.1% | <0.1% | <0.1% | <0.1% |
| $FeCl_3$+Mont | <0.1% | 0.11% | <0.1% | <0.1% | <0.1% | <0.1% | <0.1% | <0.1% |
| $FeCl_3$+Org | <0.1% | 0.12% | <0.1% | <0.1% | <0.1% | <0.1% | <0.1% | <0.1% |
| Alum | 55.0% | 60.0% | 34.5% | 36.5% | 22.0% | 20.5% | 14.0% | 22.0% |
| Alum+Kaol | 41.3% | 53.1% | 33.8% | 38.8% | 18.1% | 15.6% | 15.6% | 21.9% |
| Alum+Mont | 43.3% | 48.7% | 29.3% | 34.0% | 17.3% | 17.3% | 20.0% | 22.0% |
| Alum+Org | 75.0% | 75.0% | 49.4% | 51.3% | 31.3% | 31.3% | 20.0% | 26.9% |
| $Ca(OH)_2$+$Na_2CO_3$ | 69.8% | 72.1% | 60.5% | 60.5% | 37.2% | 41.9% | 23.3% | 25.6% |

**Table 2.11. Absolute Amounts of Arsenate Released at Various pH and Leaching Times (units: mg As/kg dry solid)**

| Sample | pH 4 | | pH 6 | | pH 8 | | pH 10 | |
|---|---|---|---|---|---|---|---|---|
| | 8 hr | 24 hr | 8 hr | 24 hr | 8 hr | 24 hr | 8 hr | 24 hr |
| $FeCl_3$ | BDL | BDL | BDL | BDL | 1.2 | 3.0 | 53 | 83 |
| $FeCl_3$+Kaol | BDL | BDL | BDL | BDL | 0.75 | 1.1 | 56 | 110 |
| $FeCl_3$+Mont | BDL | BDL | BDL | BDL | 2.7 | 5.4 | 57 | 140 |
| $FeCl_3$+Org | BDL | BDL | BDL | BDL | 3.2 | 9.0 | 60 | 110 |
| Alum | 3.3 | 6.9 | 1.3 | 1.3 | 0.65 | 0.65 | 63 | 110 |
| Alum+Kaol | 6.6 | 9.8 | 0.61 | 0.61 | 1.5 | 0.61 | 63 | 150 |
| Alum+Mont | 5.0 | 7.5 | 1.2 | 0.53 | 1.5 | 0.78 | 57 | 140 |
| Alum+Org | 2.0 | 3.6 | 0.68 | 0.68 | 1.2 | 0.68 | 93 | 250 |
| $Ca(OH)_2$+$Na_2CO_3$ | 370 | 370 | 43 | 39 | 0.16 | BDL | BDL | BDL |

**Table 2.12. Relative Amounts of Arsenate Released at Various pH and Leaching Times (unit: %)**

| Sample | pH 4 | | pH 6 | | pH 8 | | pH 10 | |
|---|---|---|---|---|---|---|---|---|
| | 8 hr | 24 hr | 8 hr | 24 hr | 8 hr | 24 hr | 8 hr | 24 hr |
| $FeCl_3$ | BDL | BDL | BDL | BDL | <0.1% | <0.1% | <0.1% | <0.1% |
| $FeCl_3$+Kaol | BDL | BDL | BDL | BDL | <0.1% | <0.1% | <0.1% | <0.1% |
| $FeCl_3$+Mont | BDL | BDL | BDL | BDL | <0.1% | <0.1% | <0.1% | <0.1% |
| $FeCl_3$+Org | BDL | BDL | BDL | BDL | <0.1% | <0.1% | <0.1% | <0.1% |
| Alum | <0.1% | <0.1% | <0.1% | <0.1% | <0.1% | <0.1% | <0.1% | <0.1% |
| Alum+Kaol | <0.1% | <0.1% | <0.1% | <0.1% | <0.1% | <0.1% | <0.1% | <0.1% |
| Alum+Mont | <0.1% | <0.1% | <0.1% | <0.1% | <0.1% | <0.1% | <0.1% | <0.1% |
| Alum+Org | <0.1% | <0.1% | <0.1% | <0.1% | <0.1% | <0.1% | <0.1% | 0.13% |
| $Ca(OH)_2$+$Na_2CO_3$ | 46.8% | 46.8% | 5.44% | 4.94% | <0.1% | BDL | BDL | BDL |

*BDL – Below Detection Limit

## Arsenic Desorption Kinetics in the Presence of Background Anions

Desorption in the presence of background anions produced some very interesting results. The results of desorption experiments conducted in the presence of 1.33 mM phosphate, sulfate and chloride appear in Tables 2.13 to 2.24. Graphical representation of the kinetic data is displayed in Figures 2.3 to 2.8. Prior to this study, the answer to the arsenic residuals problems was to use ferric chloride, oxidize arsenite to arsenate, and maintain pH below 10. However, the data of Table 2.15 shows that arsenate was significantly desorbed from the ferric chloride coagulated residuals in the presence of phosphate at a concentration of 1.33 mM. This concentration is much higher than would be expected in water supplies, but the impact of lower phosphate concentrations is not known. Arsenite was also desorbed from the iron oxy-hydroxide surface in the presence of phosphate, though not to the extent as arsenate. Arsenite was desorbed from alum-generated residuals in all environments, and thus aluminum oxy-hydroxide residuals should be avoided for arsenite removal. Arsenate likely forms an inner sphere complex with the iron surface and thus phosphate, having a similar chemistry to arsenic, readily competes for bonding sites on the iron oxy-hydroxide surface. More substantial discussions of these phenomena are found in Chapters 5 and 6.

**Table 2.13. Absolute Amounts of Arsenite Released in Presence of 1.33 mM Phosphate (unit: mg As/kg dry solids)**

| Sludge Type | Time (hr) | | | | | | |
|---|---|---|---|---|---|---|---|
| | 0.5 | 1 | 2 | 4 | 8 | 12 | 24 |
| $FeCl_3$ | 24 | 34 | 20 | 41 | 20 | 21 | 23 |
| $FeCl_3$+Kaol | 27 | 31 | 29 | 27 | 31 | 27 | 27 |
| $FeCl_3$+Mont | 29 | 29 | 27 | 29 | 26 | 29 | 26 |
| $FeCl_3$+Org | 40 | 37 | 37 | 33 | 35 | 30 | 31 |
| Alum | 4,800 | 4,900 | 5,300 | 5,100 | 5,200 | 5,100 | 5,100 |
| Alum+Kaol | 3,200 | 3,300 | 3,300 | 3,400 | 3,500 | 3,400 | 3,300 |
| Alum+Mont | 2,000 | 2,000 | 1,900 | 2,000 | 1,900 | 2,000 | 1,900 |
| Alum+Org | 2,500 | 2,500 | 2,700 | 2,400 | 2,500 | 2,200 | 2,600 |
| $CaCO_3$ | 56 | 62 | 62 | 77 | 59 | 53 | 56 |

**Table 2.14. Relative Amounts of Arsenite Released in Presence of 1.33 mM Phosphate (unit: %)**

| Sludge Type | Time (hr) | | | | | | |
|---|---|---|---|---|---|---|---|
| | 0.5 | 1 | 2 | 4 | 8 | 12 | 24 |
| $FeCl_3$ | <0.1% | 0.10% | <0.1% | 0.12% | <0.1% | <0.1% | <0.1% |
| $FeCl_3$+Kaol | <0.1% | 0.10% | <0.1% | <0.1% | 0.10% | <0.1% | <0.1% |
| $FeCl_3$+Mont | <0.1% | <0.1% | <0.1% | <0.1% | <0.1% | <0.1% | <0.1% |
| $FeCl_3$+Org | 0.13% | 0.12% | 0.12% | 0.11% | 0.12% | 0.10% | 0.10% |
| Alum | 24.0% | 24.5% | 26.5% | 25.5% | 26.0% | 25.5% | 25.5% |
| Alum+Kaol | 20.0% | 20.6% | 20.6% | 21.3% | 21.9% | 21.3% | 20.6% |
| Alum+Mont | 13.3% | 13.3% | 12.7% | 13.3% | 12.7% | 13.3% | 12.7% |
| Alum+Org | 15.6% | 15.6% | 16.9% | 15.0% | 15.6% | 13.8% | 16.3% |
| $CaCO_3$ | 13.0% | 14.4% | 14.4% | 17.9% | 13.7% | 12.3% | 13.0% |

**Table 2.15. Absolute Amounts of Arsenate Released in Presence of 1.33 mM Phosphate (unit: mg As/kg dry solids)**

| Sludge Type | Time (hr) | | | | | | |
|---|---|---|---|---|---|---|---|
| | 0.5 | 1 | 2 | 4 | 8 | 12 | 24 |
| $FeCl_3$ | 220 | 120 | 52 | 50 | 40 | 36 | 31 |
| $FeCl_3$+Kaol | 43 | 41 | 41 | 38 | 36 | 28 | 27 |
| $FeCl_3$+Mont | 100 | 96 | 89 | 86 | 81 | 64 | 61 |
| $FeCl_3$+Org | 220 | 210 | 200 | 190 | 160 | 150 | 130 |
| Alum | 2.7 | 2.4 | 2.3 | 2.1 | 1.9 | 1.9 | 1.6 |
| Alum+Kaol | 4.1 | 3.2 | 2.7 | 1.9 | 1.6 | 1.2 | 1.2 |
| Alum+Mont | 4.1 | 3.9 | 3.9 | 3 | 2.7 | 2.5 | 2.1 |
| Alum+Org | 6.7 | 4.6 | 3.6 | 3.2 | 2.9 | 2.7 | 2.3 |
| $CaCO_3$ | 0.51 | 0.11 | 0.11 | 0.11 | 0.11 | 0.31 | 0.11 |

**Table 2.16. Relative Amounts of Arsenate Released in Presence of 1.33 mM Phosphate (unit: %)**

| Sludge Type | Time (hr) | | | | | | |
|---|---|---|---|---|---|---|---|
| | 0.5 | 1 | 2 | 4 | 8 | 12 | 24 |
| $FeCl_3$ | <0.1% | <0.1% | <0.1% | <0.1% | <0.1% | <0.1% | <0.1% |
| $FeCl_3$+Kaol | <0.1% | <0.1% | <0.1% | <0.1% | <0.1% | <0.1% | <0.1% |
| $FeCl_3$+Mont | <0.1% | <0.1% | <0.1% | <0.1% | <0.1% | <0.1% | <0.1% |
| $FeCl_3$+Org | 0.10% | <0.1% | <0.1% | <0.1% | <0.1% | <0.1% | <0.1% |
| Alum | <0.1% | <0.1% | <0.1% | <0.1% | <0.1% | <0.1% | <0.1% |
| Alum+Kaol | <0.1% | <0.1% | <0.1% | <0.1% | <0.1% | <0.1% | <0.1% |
| Alum+Mont | <0.1% | <0.1% | <0.1% | <0.1% | <0.1% | <0.1% | <0.1% |
| Alum+Org | <0.1% | <0.1% | <0.1% | <0.1% | <0.1% | <0.1% | <0.1% |
| $CaCO_3$ | <0.1% | <0.1% | <0.1% | <0.1% | <0.1% | <0.1% | <0.1% |

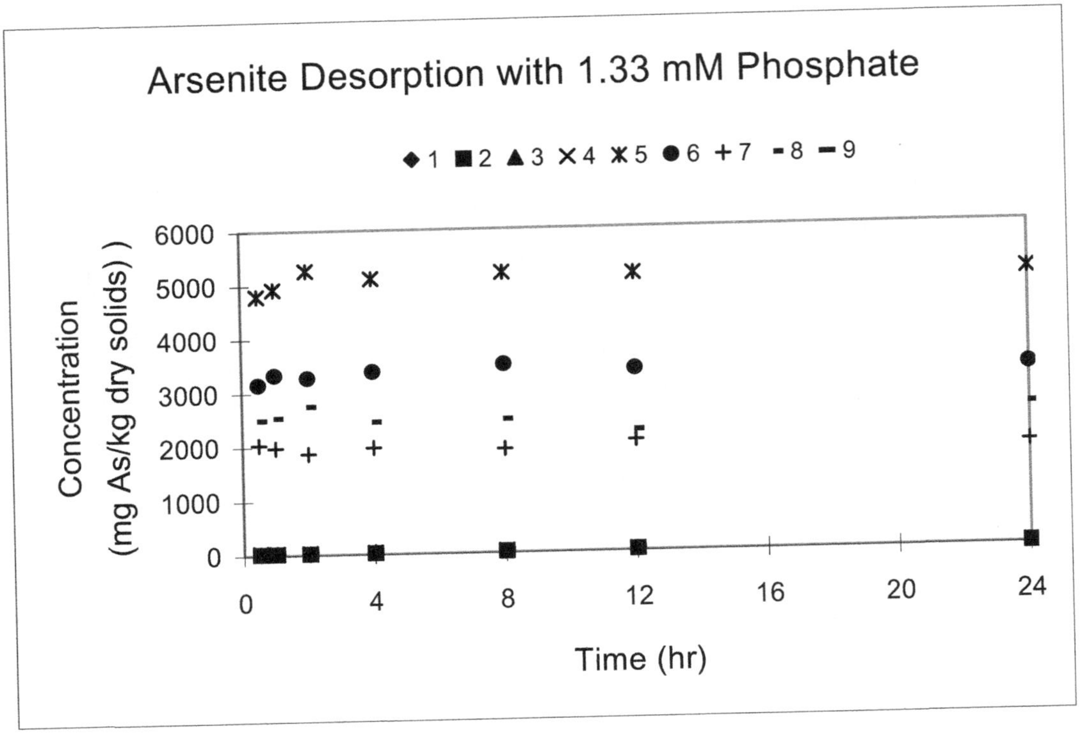


1-FeCl3, 2-FeCl3+Kaol, 3-FeCl3+Mont, 4-FeCl3+Org, 5-Alum, 6-Alum+Kaol, 7-Alum+Mont, 8-Alum+Org, 9-CaCO3

**Figure 2.3. Arsenite Desorption With 1.33 mM Phosphate at pH 7.0**

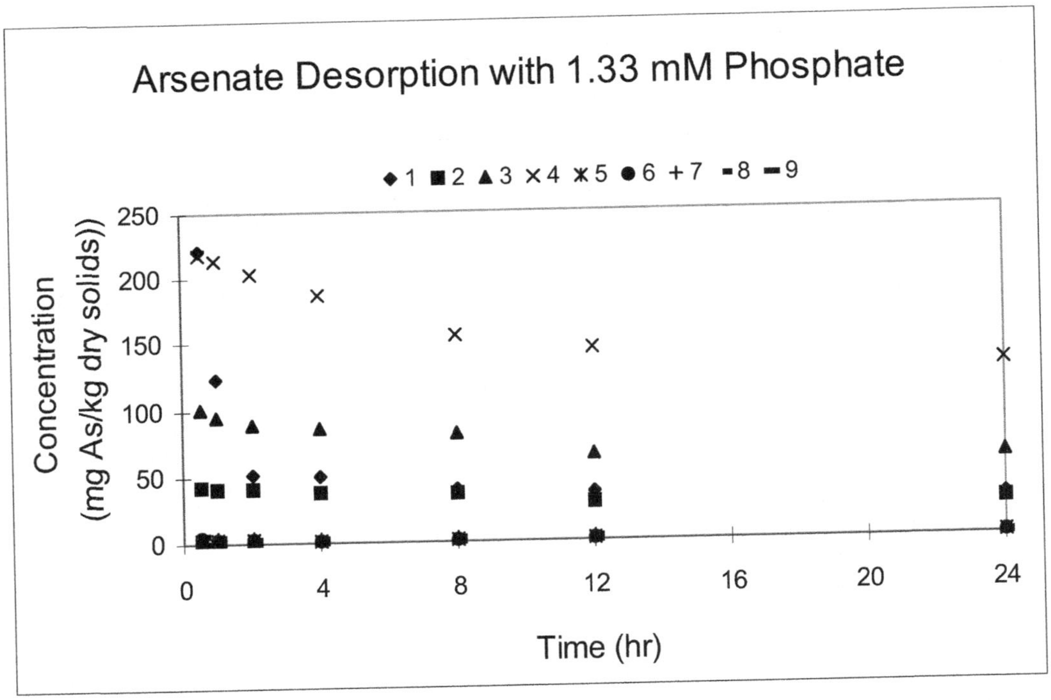


1-FeCl3, 2-FeCl3+Kaol, 3-FeCl3+Mont, 4-FeCl3+Org, 5-Alum, 6-Alum+Kaol, 7-Alum+Mont, 8-Alum+Org, 9-CaCO3

**Figure 2.4. Arsenate Desorption With 1.33 mM Phosphate at pH 7.0**

**Table 2.17. Absolute Amounts of Arsenite Released in Presence of 1.33 mM Sulfate (unit: mg As/kg dry solids)**

| Sludge Type | Time (hr) | | | | | | |
|---|---|---|---|---|---|---|---|
| | 0.5 | 1 | 2 | 4 | 8 | 12 | 24 |
| $FeCl_3$ | 7.9 | 9 | 8.9 | 9.6 | 10 | 9.6 | 9.5 |
| $FeCl_3$+Kaol | 8.7 | 10 | 10 | 11 | 12 | 11 | 11 |
| $FeCl_3$+Mont | 14 | 11 | 12 | 11 | 12 | 12 | 13 |
| $FeCl_3$+Org | 12 | 13 | 13 | 13 | 14 | 13 | 13 |
| Alum | 4,600 | 4,800 | 4,900 | 5,200 | 5,400 | 5,400 | 5,400 |
| Alum+Kaol | 3,500 | 3,600 | 3,700 | 4,000 | 4,000 | 3,900 | 4,200 |
| Alum+Mont | 2,300 | 2,300 | 2,400 | 2,500 | 2,600 | 2,500 | 2,500 |
| Alum+Org | 3,000 | 3,300 | 3,300 | 3,300 | 3,500 | 3,300 | 3,500 |
| $CaCO_3$ | 35 | 44 | 52 | 52 | 48 | 56 | 60 |

**Table 2.18. Relative Amounts of Arsenite Released in Presence of 1.33 mM Sulfate (unit: %)**

| Sludge Type | Time (hr) | | | | | | |
|---|---|---|---|---|---|---|---|
| | 0.5 | 1 | 2 | 4 | 8 | 12 | 24 |
| $FeCl_3$ | <0.1% | <0.1% | <0.1% | <0.1% | <0.1% | <0.1% | <0.1% |
| $FeCl_3$+Kaol | <0.1% | <0.1% | <0.1% | <0.1% | <0.1% | <0.1% | <0.1% |
| $FeCl_3$+Mont | <0.1% | <0.1% | <0.1% | <0.1% | <0.1% | <0.1% | <0.1% |
| $FeCl_3$+Org | <0.1% | <0.1% | <0.1% | <0.1% | <0.1% | <0.1% | <0.1% |
| Alum | 23.0% | 24.0% | 24.5% | 26.0% | 27.0% | 27.0% | 27.0% |
| Alum+Kaol | 21.9% | 22.5% | 23.1% | 25.0% | 25.0% | 24.4% | 26.3% |
| Alum+Mont | 15.3% | 15.3% | 16.0% | 16.7% | 17.3% | 16.7% | 16.7% |
| Alum+Org | 18.8% | 20.6% | 20.6% | 20.6% | 21.9% | 20.6% | 21.9% |
| $CaCO_3$ | 8.14% | 10.2% | 12.1% | 12.1% | 11.2% | 13.0% | 14.0% |

**Table 2.19. Absolute Amounts of Arsenate Released in Presence of 1.33 mM Sulfate (unit: mg As/kg dry solids)**

| Sludge Type | Time (hr) | | | | | | |
|---|---|---|---|---|---|---|---|
| | 0.5 | 1 | 2 | 4 | 8 | 12 | 24 |
| $FeCl_3$ | 0.87 | 0.58 | 0.58 | 0.58 | 0.58 | 0.87 | 0.58 |
| $FeCl_3$+Kaol | 0.53 | 0.26 | 0.26 | 0.26 | 0.26 | 0.26 | 0.26 |
| $FeCl_3$+Mont | 1.3 | 1.3 | 1.5 | 1.5 | 1.8 | 2 | 1.5 |
| $FeCl_3$+Org | 3.8 | 4.1 | 3.8 | 3.8 | 3.5 | 3.5 | 3.2 |
| Alum | 0.69 | 0.46 | 0.69 | 0.69 | 0.69 | 0.46 | 1.2 |
| Alum+Kaol | 2.4 | 0.86 | 1.1 | 1.1 | 1.1 | 1.3 | 1.1 |
| Alum+Mont | 0.92 | 0.55 | 1.1 | 1.1 | 0.92 | 1.1 | 1.9 |
| Alum+Org | 1.2 | 1.2 | 1.2 | 1.2 | 1.7 | 0.96 | 0.84 |
| $CaCO_3$ | 1.4 | 2.1 | 1.8 | 0.93 | 1.1 | 1 | 1.1 |

**Table 2.20. Relative Amounts of Arsenate Released in Presence of 1.33 mM Sulfate (unit: %)**

| Sludge Type | Time (hr) | | | | | | |
|---|---|---|---|---|---|---|---|
| | 0.5 | 1 | 2 | 4 | 8 | 12 | 24 |
| $FeCl_3$ | <0.1% | <0.1% | <0.1% | <0.1% | <0.1% | <0.1% | <0.1% |
| $FeCl_3$+Kaol | <0.1% | <0.1% | <0.1% | <0.1% | <0.1% | <0.1% | <0.1% |
| $FeCl_3$+Mont | <0.1% | <0.1% | <0.1% | <0.1% | <0.1% | <0.1% | <0.1% |
| $FeCl_3$+Org | <0.1% | <0.1% | <0.1% | <0.1% | <0.1% | <0.1% | <0.1% |
| Alum | <0.1% | <0.1% | <0.1% | <0.1% | <0.1% | <0.1% | <0.1% |
| Alum+Kaol | <0.1% | <0.1% | <0.1% | <0.1% | <0.1% | <0.1% | <0.1% |
| Alum+Mont | <0.1% | <0.1% | <0.1% | <0.1% | <0.1% | <0.1% | <0.1% |
| Alum+Org | <0.1% | <0.1% | <0.1% | <0.1% | <0.1% | <0.1% | <0.1% |
| $CaCO_3$ | 0.18% | 0.27% | 0.23% | 0.12% | 0.14% | 0.13% | 0.14% |

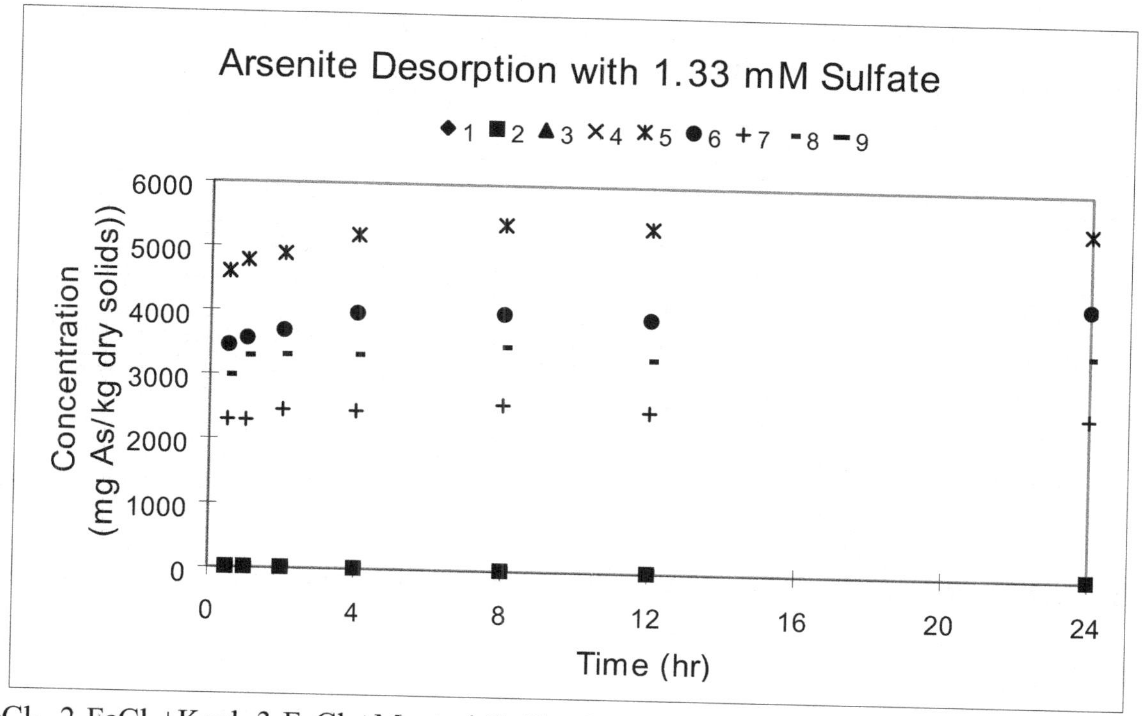


1-$FeCl_3$, 2-$FeCl_3$+Kaol, 3-$FeCl_3$+Mont, 4-$FeCl_3$+Org, 5-Alum, 6-Alum+Kaol, 7-Alum+Mont, 8-Alum+Org, 9-$CaCO_3$

**Figure 2.5. Arsenite Desorption With 1.33 mM Sulfate at pH 7.0**

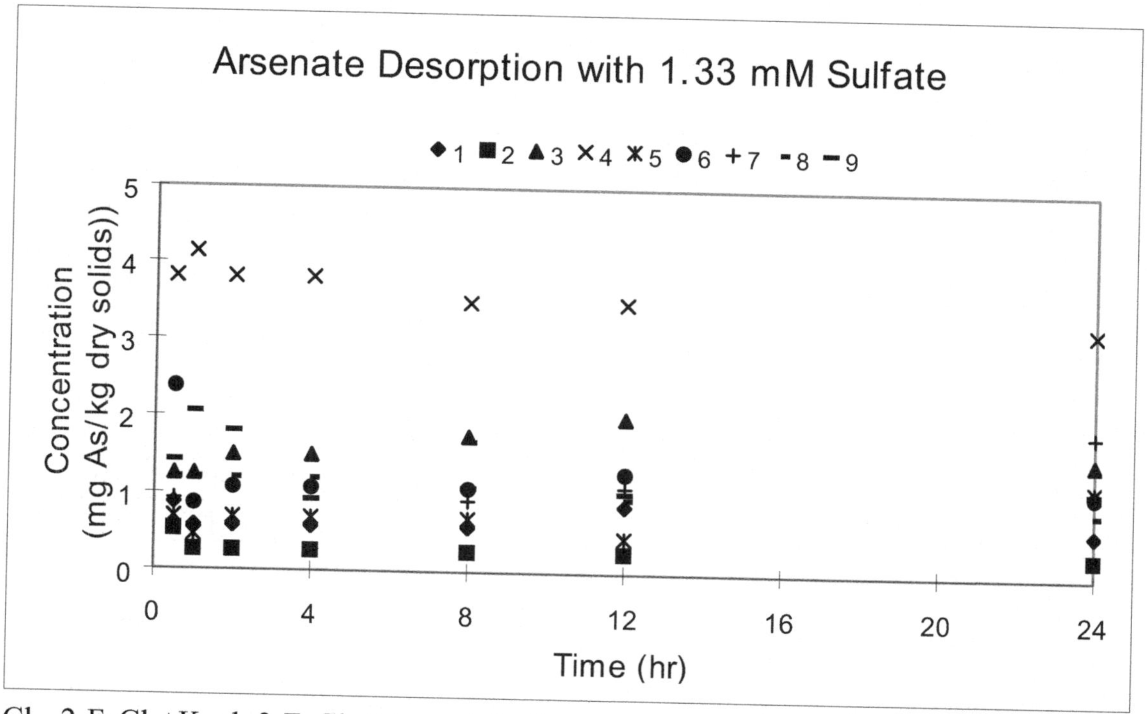


1-$FeCl_3$, 2-$FeCl_3$+Kaol, 3-$FeCl_3$+Mont, 4-$FeCl_3$+Org, 5-Alum, 6-Alum+Kaol, 7-Alum+Mont, 8-Alum+Org, 9-$CaCO_3$

**Figure 2.6. Arsenate Desorption With 1.33 mM Sulfate at pH 7.0**

**Table 2.21. Absolute Amounts of Arsenite Released in Presence of 1.33 mM Chloride (unit: mg As/kg dry solids)**

| Sludge Type | Time (hr) | | | | | | |
|---|---|---|---|---|---|---|---|
| | 0.5 | 1 | 2 | 4 | 8 | 12 | 24 |
| $FeCl_3$ | 12 | 12 | 12 | 12 | 13 | 12 | 12 |
| $FeCl_3$+Kaol | 12 | 13 | 13 | 13 | 14 | 14 | 12 |
| $FeCl_3$+Mont | 14 | 13 | 14 | 13 | 13 | 14 | 14 |
| $FeCl_3$+Org | 14 | 14 | 15 | 14 | 14 | 15 | 15 |
| Alum | 4,700 | 4,800 | 4,900 | 5,000 | 4,900 | 5,000 | 5,000 |
| Alum+Kaol | 2,800 | 3,000 | 3,000 | 3,100 | 3,100 | 3,200 | 3,300 |
| Alum+Mont | 1,900 | 1,800 | 1,800 | 1,900 | 2,100 | 1,900 | 2,100 |
| Alum+Org | 1,800 | 2,200 | 2,400 | 2,400 | 2,700 | 2,800 | 2,600 |
| $CaCO_3$ | 35 | 38 | 40 | 43 | 45 | 48 | 51 |

**Table 2.22. Relative Amounts of Arsenite Released in Presence of 1.33 mM Chloride (unit: %)**

| Sludge Type | Time (hr) | | | | | | |
|---|---|---|---|---|---|---|---|
| | 0.5 | 1 | 2 | 4 | 8 | 12 | 24 |
| $FeCl_3$ | <0.1% | <0.1% | <0.1% | <0.1% | <0.1% | <0.1% | <0.1% |
| $FeCl_3$+Kaol | <0.1% | <0.1% | <0.1% | <0.1% | <0.1% | <0.1% | <0.1% |
| $FeCl_3$+Mont | <0.1% | <0.1% | <0.1% | <0.1% | <0.1% | <0.1% | <0.1% |
| $FeCl_3$+Org | <0.1% | <0.1% | <0.1% | <0.1% | <0.1% | <0.1% | <0.1% |
| Alum | 23.5% | 24.0% | 24.5% | 25.0% | 24.5% | 25.0% | 25.0% |
| Alum+Kaol | 17.5% | 18.8% | 18.8% | 19.4% | 19.4% | 20.0% | 20.6% |
| Alum+Mont | 12.7% | 12.0% | 12.0% | 12.7% | 14.0% | 12.7% | 14.0% |
| Alum+Org | 11.3% | 13.8% | 15.0% | 15.0% | 16.9% | 17.5% | 16.3% |
| $CaCO_3$ | 8.14% | 8.84% | 9.30% | 10.0% | 10.5% | 11.2% | 11.9% |

**Table 2.23. Absolute Amounts of Arsenate Released in Presence of 1.33 mM Chloride (unit: mg As/kg dry solids)**

| Sludge Type | Time (hr) | | | | | | |
|---|---|---|---|---|---|---|---|
| | 0.5 | 1 | 2 | 4 | 8 | 12 | 24 |
| $FeCl_3$ | 0.4 | 0.88 | 0.72 | 0.72 | 0.4 | 0.4 | 0.72 |
| $FeCl_3$+Kaol | BDL | BDL | BDL | BDL | BDL | BDL | BDL |
| $FeCl_3$+Mont | 1.2 | 1.5 | 1.5 | 1.2 | 1.2 | 1.7 | 1 |
| $FeCl_3$+Org | 4.6 | 3.9 | 4.3 | 5 | 4.6 | 4.6 | 3.9 |
| Alum | 0.7 | 1.2 | 0.82 | 0.82 | 0.82 | 0.95 | 0.7 |
| Alum+Kaol | 0.65 | 0.77 | 0.77 | 1 | 0.77 | 0.77 | 0.77 |
| Alum+Mont | 0.88 | 0.77 | 0.67 | 0.67 | 0.67 | 0.67 | 0.88 |
| Alum+Org | 1.1 | 1.1 | 0.99 | 1.4 | 1.1 | 1.4 | 1.1 |
| $CaCO_3$ | 2.3 | 1.9 | 1.6 | 1.7 | 1.7 | 1.6 | 2 |

**Table 2.24. Relative Amounts of Arsenate Released in Presence of 1.33 mM Chloride (unit: %)**

| Sludge Type | Time (hr) | | | | | | |
|---|---|---|---|---|---|---|---|
| | 0.5 | 1 | 2 | 4 | 8 | 12 | 24 |
| $FeCl_3$ | <0.1% | <0.1% | <0.1% | <0.1% | <0.1% | <0.1% | <0.1% |
| $FeCl_3$+Kaol | BDL | BDL | BDL | BDL | BDL | BDL | BDL |
| $FeCl_3$+Mont | <0.1% | <0.1% | <0.1% | <0.1% | <0.1% | <0.1% | <0.1% |
| $FeCl_3$+Org | <0.1% | <0.1% | <0.1% | <0.1% | <0.1% | <0.1% | <0.1% |
| Alum | <0.1% | <0.1% | <0.1% | <0.1% | <0.1% | <0.1% | <0.1% |
| Alum+Kaol | <0.1% | <0.1% | <0.1% | <0.1% | <0.1% | <0.1% | <0.1% |
| Alum+Mont | <0.1% | <0.1% | <0.1% | <0.1% | <0.1% | <0.1% | <0.1% |
| Alum+Org | <0.1% | <0.1% | <0.1% | <0.1% | <0.1% | <0.1% | <0.1% |
| $CaCO_3$ | 0.29% | 0.24% | 0.20% | 0.22% | 0.22% | 0.20% | 0.25% |

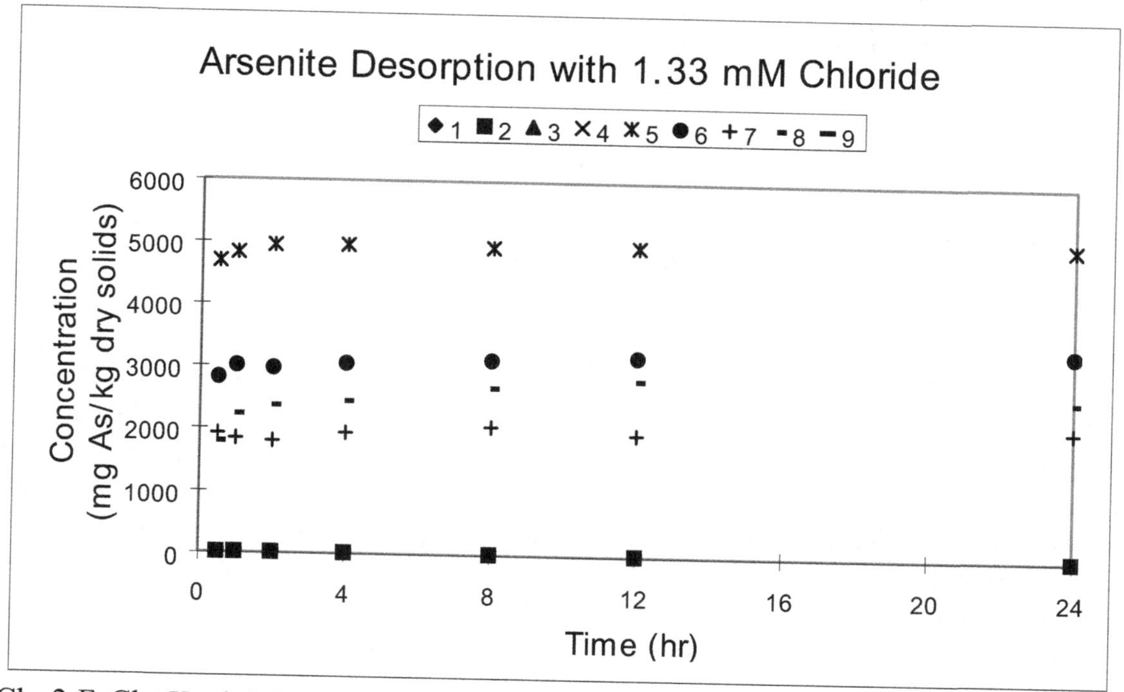


1-$FeCl_3$, 2-$FeCl_3$+Kaol, 3-$FeCl_3$+Mont, 4-$FeCl_3$+Org, 5-Alum, 6-Alum+Kaol, 7-Alum+ Mont, 8--Alum+Org, 9-$CaCO_3$

**Figure 2.7. Arsenite Desorption With 1.33 mM Chloride at pH 7.0**

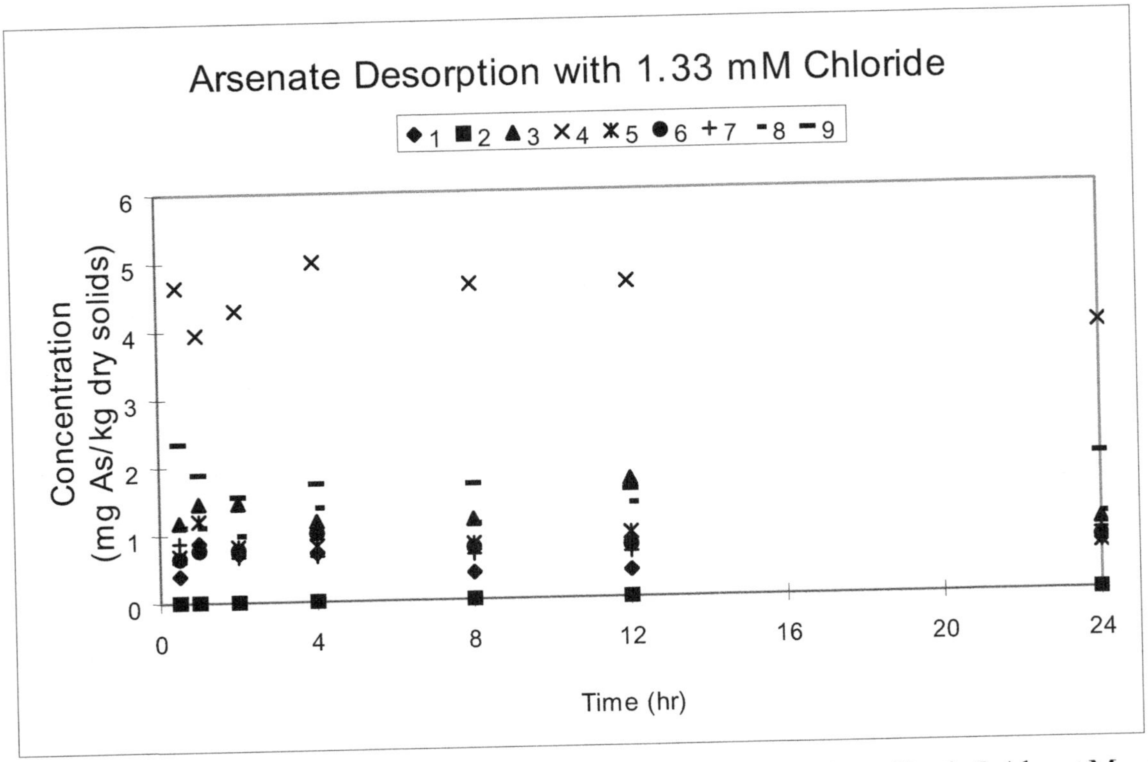


1-FeCl3, 2-FeCl3+Kaol, 3-FeCl3+Mont, 4-FeCl3+Org, 5-Alum, 6-Alum+Kaol, 7-Alum+Mont, 8-Alum+Org, 9-CaCO3

**Figure 2.8. Arsenate Desorption With 1.33 mM Chloride at pH 7.0**

One additional item to consider in the leaching studies is the strong ability of the residuals to buffer or alter pH. This topic is further discussed in the following Chapter. In addition to the measurement and adjustment of solution pH at the beginning of each test, pH was also measured at the end of each desorption test to check for possible changes. As expected, both the iron and aluminum residuals tended to move the pH in a neutral direction, i.e. towards 7.0. This pH shift is due to the exchange of protons and hydroxides by the oxy-hydroxides with the solution, depending on pH, and thus their behavior as strong buffers. Calcium carbonate on the other hand tended to raise pH, as can be seen in the Tables 2.25 and 2.26.

**Table 2.25. Arsenite Solution pH Before and After Leaching**

| Sample | pH 4 | | pH 6 | | pH 8 | | pH 10 | |
|---|---|---|---|---|---|---|---|---|
| | Initial PH | Final pH* | Initial pH | Final pH | Initial pH | Final pH | Initial pH | Final pH |
| $FeCl_3$ | 4.1 | 4.7 | 6.1 | 6.1 | 8.0 | 8.3 | 10.0 | 9.8 |
| $FeCl_3$+Kaol | 4.0 | 4.4 | 6.0 | 6.1 | 8.1 | 8.0 | 10.1 | 10.1 |
| $FeCl_3$+Mont | 4.0 | 4.5 | 6.0 | 6.1 | 8.1 | 7.7 | 10.0 | 9.7 |
| $FeCl_3$+Org | 4.1 | 4.5 | 6.0 | 6.0 | 8.1 | 7.7 | 10.1 | 9.5 |
| Alum | 4.1 | 4.4 | 6.1 | 5.2 | 8.0 | 7.5 | 10.1 | 9.8 |
| Alum+Kaol | 4.1 | 4.6 | 5.9 | 6.0 | 8.0 | 7.6 | 10.1 | 9.8 |
| Alum+Mont | 4.0 | 4.6 | 6.0 | 6.2 | 8.1 | 7.6 | 10.0 | 10.3 |
| Alum+Org | 4.0 | 4.5 | 5.9 | 6.2 | 8.0 | 7.4 | 10.1 | 9.7 |
| $CaCO_3$ | 4.1 | 5.7 | 5.9 | 6.3 | 7.9 | 8.8 | 10.1 | 10.8 |

*Final pH means a pH value after 24 hours of reaction

**Table 2.26. Arsenate Solution pH Before and After Leaching**

| Sample | pH 4 | | pH 6 | | pH 8 | | pH 10 | |
|---|---|---|---|---|---|---|---|---|
| | Initial PH | Final pH* | Initial pH | Final pH | Initial pH | Final pH | Initial pH | Final pH |
| $FeCl_3$ | 3.9 | 3.9 | 6.0 | 5.9 | 8.1 | 7.9 | 10.1 | 9.5 |
| $FeCl_3$+Kaol | 4.0 | 4.2 | 5.9 | 5.9 | 8.0 | 7.7 | 10.0 | 9.8 |
| $FeCl_3$+Mont | 3.9 | 4.3 | 6.0 | 5.9 | 8.1 | 8.0 | 9.9 | 9.9 |
| $FeCl_3$+Org | 4.0 | 4.3 | 6.1 | 6.4 | 8.0 | 7.9 | 10.0 | 9.6 |
| Alum | 3.9 | 4.6 | 6.0 | 6.0 | 8.0 | 7.4 | 10.0 | 9.3 |
| Alum+Kaol | 3.9 | 4.5 | 5.9 | 6.0 | 8.1 | 7.5 | 10.1 | 9.5 |
| Alum+Mont | 3.9 | 4.6 | 5.9 | 6.1 | 7.9 | 7.3 | 10.0 | 9.3 |
| Alum+Org | 3.9 | 4.6 | 5.9 | 6.1 | 8.1 | 7.3 | 10.0 | 9.7 |
| $CaCO_3$ | 3.9 | 3.9 | 6.0 | 6.3 | 8.1 | 10.5 | 10.0 | 10.9 |

## SUMMARY

Experimental results presented in this chapter showed that lime-soda residuals have higher dry solids concentrations than those based on aluminum or iron. Arsenate was removed from solution much more effectively than arsenite by both iron and aluminum solids. Aluminum residuals were less effective in removing arsenic than iron residuals, particularly when arsenite was being removed. Addition of clays (Kaolinite, Montmorillonite) did not improve the residuals ability to remove arsenic.

The amount of arsenite released from the residuals under conditions of the TCLP increased with time and reached a relatively constant value after about 10 hours. The amount of arsenate leached decreased a little over time.

Arsenite and arsenate behaved differently in experiments investigating the effects of pH on leaching. Arsenite leaching increased as pH was decreased and arsenate leaching increased as pH was increased. More arsenic was released from aluminum residuals than from iron residuals and much more arsenite was released than arsenate from both types.

Presence of 1.33 mM phosphate resulted in increased release from iron residuals of arsenate and, to a lesser extent, arsenite. The presence of 1.33 mM sulfate or chloride had little effect on arsenic release. The effects of anions on release of arsenic from aluminum residuals were difficult to measure, due the generally high amounts released without addition of competing anions.

Residuals can have an effect on pH. Aluminum and iron residuals tended to raise or lower pH in the direction of pH 7, while lime-soda residuals tended to raise pH under all conditions studied.

Results from this phase of the project indicate that alum is not a very desirable sorbent for removing arsenite due to its weak ability to remove and retain arsenite. The oxy-hydroxides of iron appear to be very stable; however, phosphate is a major problem. Calcium could be the key to forming extremely stable residuals over a broad pH range and variable ionic environment.

# CHAPTER 3: BUFFER CAPACITY OF THE RESIDUALS

## INTRODUCTION

As mentioned in the previous chapter metal oxy-hydrides have a strong tendency to buffer water pH. The significance of this behavior is that if the residuals can buffer themselves, chemical treatments to maintain environmental pH may not be necessary. Initial solution (leachate liquid) pH and the final pH after residuals were added and allowed to equilibrate for 24 hrs are shown for both arsenite and arsenate residuals in Tables 3.1 and 3.2. Both the iron and aluminum residuals tend to move the pH towards pH 7.0. This behavior is due to the oxy-hydroxide's ability to exchange protons and hydroxides with the solution, depending on pH, and thus act as strong buffers. This tendency can be characterized with chemical equilibrium models that describe the ability of the surface to exchange protons in a way that is similar to the ability of acids or bases in solution to exchange protons. These models use surface acidity constants that are analogous to solution acid disassociation constants. Calcium carbonate on the other hand tends to raise pH as can be seen in the tables. A method is needed to quantify this behavior in order to predict solution pH after the residuals have been in contact with an elevated or lowered pH environment for a period of time. More specifically, how well can a given residuals sample maintain neutral pH in the natural environment.

**Table 3.1. Arsenite Residuals pH After Addition to the Leachate Solution**

| Sludge Type | Final pH in solution after 24 hr reaction | | | |
|---|---|---|---|---|
| | Initial pH 4 | Initial pH 6 | Initial pH 8 | Initial pH 10 |
| $FeCl_3$ | 6.6 | 7.6 | 7.4 | 7.1 |
| $FeCl_3$+Kaol | 7.1 | 7.5 | 7.5 | 7.5 |
| $FeCl_3$+Mont | 6.7 | 7.0 | 7.1 | 7.1 |
| $FeCl_3$+Org | 7.27 | 7.5 | 7.6 | 7.6 |
| Alum | 6.6 | 6.8 | 6.9 | 6.8 |
| Alum+Kaol | 6.7 | 6.9 | 6.9 | 6.9 |
| Alum+Mont | 6.8 | 6.9 | 7.0 | 6.9 |
| Alum+Org | 6.6 | 6.7 | 6.8 | 6.8 |
| $CaCO_3$ | 13.4 | 13.4 | 13.4 | 13.5 |

**Table 3.2. Arsenate Residuals pH After Addition to the Leachate Solution**

| Sludge Type | Final pH in solution after 24 hr reaction | | | |
|---|---|---|---|---|
| | Initial pH 4 | Initial pH 6 | Initial pH 8 | Initial pH 10 |
| $FeCl_3$ | 7.6 | 7.8 | 7.8 | 7.9 |
| $FeCl_3$+Kaol | 7.3 | 7.3 | 7.4 | 7.5 |
| $FeCl_3$+Mont | 7.6 | 7.6 | 7.7 | 7.8 |
| $FeCl_3$+Org | 7.9 | 8.0 | 8.0 | 8.1 |
| Alum | 7.1 | 7.1 | 7.1 | 7.2 |
| Alum + Kaol | 6.9 | 7.1 | 7.1 | 7.2 |
| Alum + Mont | 7.1 | 7.1 | 7.1 | 7.1 |
| Alum + Org | 7.1 | 7.1 | 7.1 | 7.2 |
| $CaCO_3$ | 13.3 | 13.4 | 13.5 | 13.5 |

Note: Extraction solutions (DI and 10 mM NaNO3) were pre-adjusted to the desired pH values (4, 6, 8, and 10). After 24 hours reaction with each sludge sample, pH in solutions was measured.

## DETERMINING THE SURFACE ACID-BASE EQUILIBRIUM CONSTANTS

Stumm and Morgan (1996) presented a method to calculate the two surface acid-base equilibrium constants based on titration data. As an example the authors of this report used a suspension of 2.5 g of activated alumina in distilled, de-ionized water and titrated in the presence of an inert electrolyte (0.01 M $NaNO_3$) with the addition of a base, NaOH, and an acid, $HNO_3$. The electrolyte was added to ensure that the addition of the acid or base does not significantly alter the ionic strength of the solution. The titration curve obtained is shown in Figure 3.1.

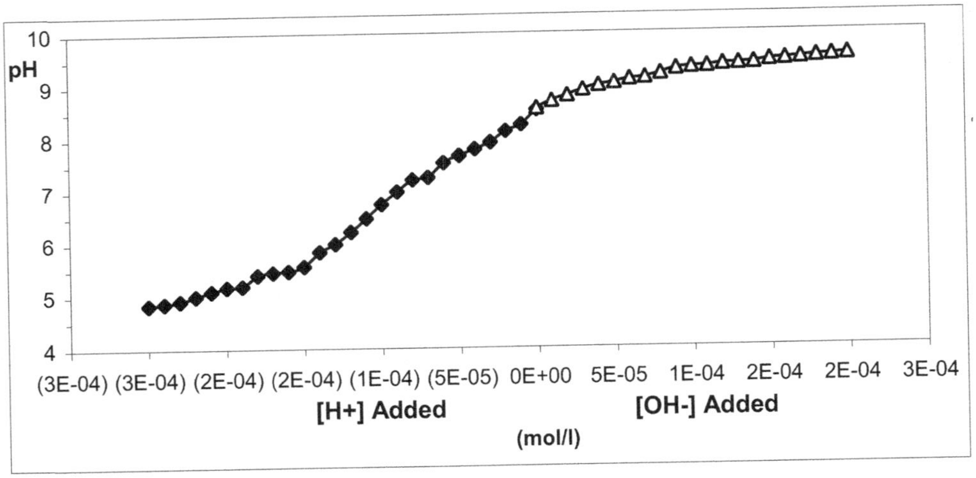


**Figure 3.1. Alumina Titration Curve**

Given the equilibrium equations for the acid-base chemistry of the surface functional groups,

$$\equiv SOH_2^+ \rightarrow \equiv SOH + H^+ ; \quad K^+ \tag{3.1}$$

$$\equiv SOH \rightarrow \equiv SO^- + H^+ ; \quad K^- \tag{3.2}$$

and applying them to the alumina complexation, the equations can be written as,

$$\equiv AlOH_2^+ \rightarrow \equiv AlOH + H^+ ; \quad K^+ \qquad (3.3)$$

$$\equiv AlOH \rightarrow \equiv AlO^- + H^+ ; \quad K^- \qquad (3.4)$$

while the surface equilibrium constants can be written as,

$$K^+ = \frac{[\equiv AlOH][H^+]}{[\equiv AlOH_2^+]} \qquad (3.5)$$

$$K^- = \frac{[\equiv AlO^-][H^+]}{[\equiv AlOH]} \qquad (3.6)$$

Considering $C_B$ and $C_A$ as the concentrations of base and acid, respectively, added per liter, the following equation can be written for every point in the titration curve:

$$C_A - C_B = [\equiv AlOH_2^+] - [\equiv AlO^-] + [H^+] - [OH^-] \qquad (3.7)$$

where [ ] indicates concentrations of solute and surface species per unit volume of solution (molar units). The total number of moles per liter of $H^+$ ions bound to $Al_2O_3$ is given by the right-hand side of equation 3.7. The portion of the charge due to the $OH^-$ or $H^+$ added (known as the mean surface charge, Q [moles/g]) can be calculated from the difference between the total acid or base added and the equilibrium $OH^-$ and $H^+$ ion concentration for a given dose of oxide, D (g/L).

$$\frac{(C_A - C_B + [OH^-] - [H^+])}{D} = \frac{[\equiv ALOH_2^+] - [\equiv AlO^-]}{D} = Q \qquad (3.8)$$

Surface charge was calculated for every point in the titration curve and the results are plotted in Figure 3.2.

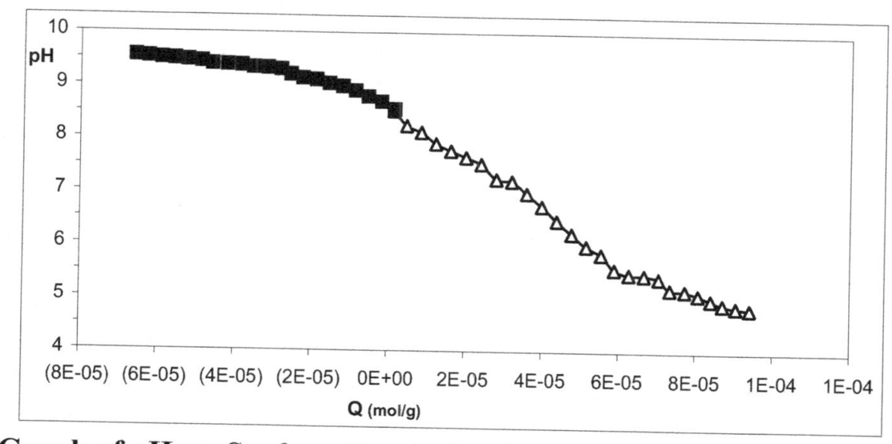


**Figure 3.2 . Graph of pH vs. Surface Charge for the Alumina Titration**

At this point there are four unknown species and four equations, thus the surface equilibrium constants can be calculated. The four unknowns are: $=AlOH_2^+$, $=AlOH$, $=AlO^-$, $H^+$. And the four equations are:

$$K^+ = \frac{[\equiv AlOH][H^+]}{[\equiv AlOH_2^+]} \qquad (3.9)$$

$$K^- = \frac{[\equiv AlO^-][H^+]}{[\equiv AlOH]} \qquad (3.10)$$

$$C_A - C_B = [\equiv AlOH_2^+] - [\equiv AlO^-] + [H^+] - [OH^-] \qquad (3.11)$$

$$[TOT\equiv AlOH] = [\equiv AlOH_2^+] + [AlOH] + [AlO^-] \qquad (3.12)$$

Where [TOT≡AlOH] is the concentration per unit mass of all alumina surface functional groups (moles/g). It has been shown that the surface area for the alumina is approximately 155 $m^2/g$. Thus, the total alumina surface functional group concentration is equal to $2.57 \times 10^{-3}$ moles/g and can be calculated using the equation,

$$[TOT \equiv AlOH] = \frac{N_t * A_S}{\text{Avogadro's number}} \qquad (3.13)$$

where $N_t$, is the total number of surface sites per unit area, $A_s$ is the alumina specific surface area, and A is Avogadro's number. Stumm and Morgan facilitate the calculation by assuming that

$$Q = [\equiv AlOH_2^+] \quad \text{for } K^+ \qquad (3.14)$$

$$Q = [\equiv AlO^-] \quad \text{for } K^- \qquad (3.15)$$

The substitution of this value into the surface equilibrium constant equations, makes it possible to calculate the equations (see Appendix A for calculations). The new equations are

$$K^+ = \frac{([TOT \equiv AlOH] - Q)[H^+]}{Q} \qquad (3.16)$$

$$K^- = \frac{Q\,[H^+]}{([TOT \equiv AlOH] - Q)} \qquad (3.17)$$

However, reactions of surface hydroxides are not the same as those of acids in solution. These equilibrium constants vary depending on the surface charge, which at the same time varies due to the exchange of hydrogen ions. The corrected values of the constants are called the intrinsic surface equilibrium constants, and they can be calculated by linear extrapolation of the log K versus surface charge curve to the zero charge condition (Stumm and Morgan, 1996), as shown below in Figure 3.3.

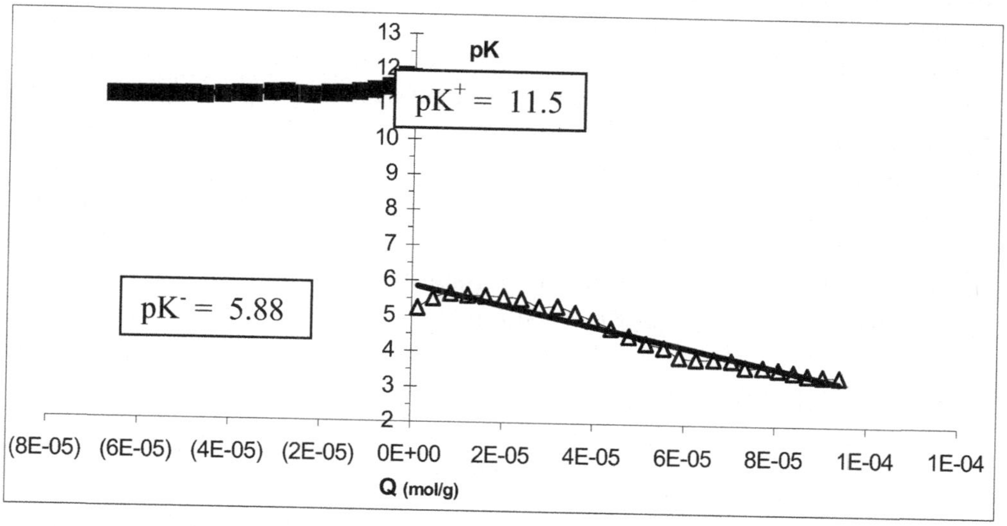


**Figure 3.3. Graph of pK Versus Surface Charge of the Alumina Titration**

While the analysis given above produces the equilibrium constant, it does rely on data that is difficult to obtain from treatment residuals, specifically the surface area and the number of charge sites. In order to assess the amount of acid or base that a buffer can absorb, the buffer capacity can be expressed as,

$$\beta = \frac{\partial C_B}{\partial pH} = -\frac{\partial C_A}{\partial pH} \tag{3.18}$$

where β is the slope of the titration curve. If β is known, the change in pH (dpH) can be obtained for any given change in proton ($dC_A$) or hydroxide ($dC_B$) concentration. Thus, it is the buffering capacity β that is desired. The buffering capacity can be calculated if the equilibrium constants are known using the relationship given as,

$$\beta = 2.3\left\{CK_1[H^+]\frac{[H^+]^2 + 4K_2[H^+] + K_1K_2}{([H^+]^2 + K_1[H^+] + K_1K_2)^2} + [H^+] + [OH^-]\right\} \tag{3.19}$$

where $K_1$ and $K_2$ are the equilibrium acid and base constants and $C = [H_2C] + [HC^-] + [C^{2-}]$

## RESULTS OF CHARACTERISTIC RESIDUALS USING THE TITRATION METHOD

Two samples of the residuals generated in Chapter 2 were titrated. One sample was the pure ferric chloride and arsenite residual, while the other was the alum/arsenite system. The acid used was 1.0 M nitric acid ($HNO_3$) and the base applied was 1.0 NaOH. The amount of residual used was 2.5 g (dry solids basis). The percentages of dry solids were determined prior to testing.

The residual sample was then added to a 1.0 L beaker and an amount of DI water added to bring the total water content up to 1.0 L. For example, the ferric residual was found to contain 10% solids, so the amount of wet solids required was 25 g and these 25 g of wet residuals contained 22.5 mL of water. Therefore, 977.5 mL of DI water was added to the beaker before the titration. Base was added in 0.1 mL increments while the beaker was stirred, and the pH was monitored and recorded once equilibrium was established. The process was repeated until the final pH was reached. An additional 2.5 g (dry basis) sample was tested by adding acid in 0.1 mL increments in an identical fashion.

The titration curve, charge balance and surface acidity equilibrium constants for the ferric chloride and arsenite residual sample appears in Figures 3.4 to 3.6. Calculation of total number of Fe surface functional groups was estimated by assuming,

$A_s$ for $Fe(OH)_3 \bullet nH_2O$ (ferrihydrite) = 306 $m^2/g$

and,

$N_t$ for $Fe(OH)_3 \bullet nH_2O$ (ferrihydrite) = 20/$nm^2$

and,

$$[TOT \equiv FeOH] = \frac{306m^2 / g \times 20 / nm^2 \times \frac{(10^9 nm)^2}{m^2}}{6.022 \times 10^{23} / mol} = 1.02 \times 10^{-2} mol / g$$

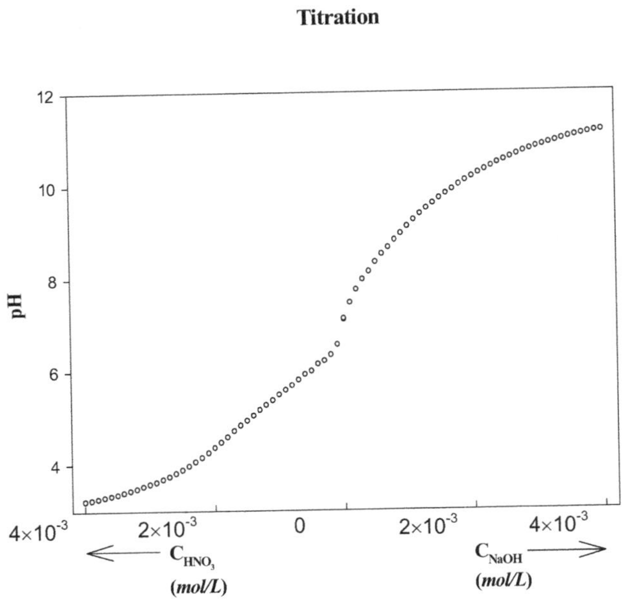


**Figure 3.4. Titration Curve for the Ferric Chloride Residuals Sample**

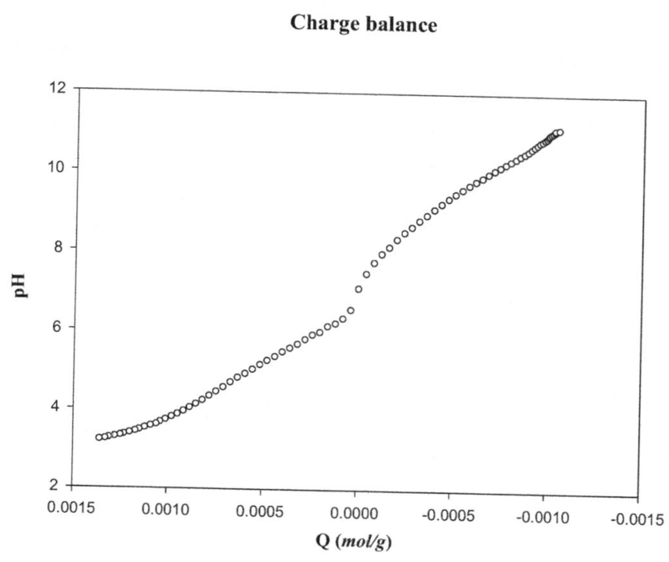


**Figure 3.5. Charge Balance Curve for the Ferric Chloride Residuals Sample**

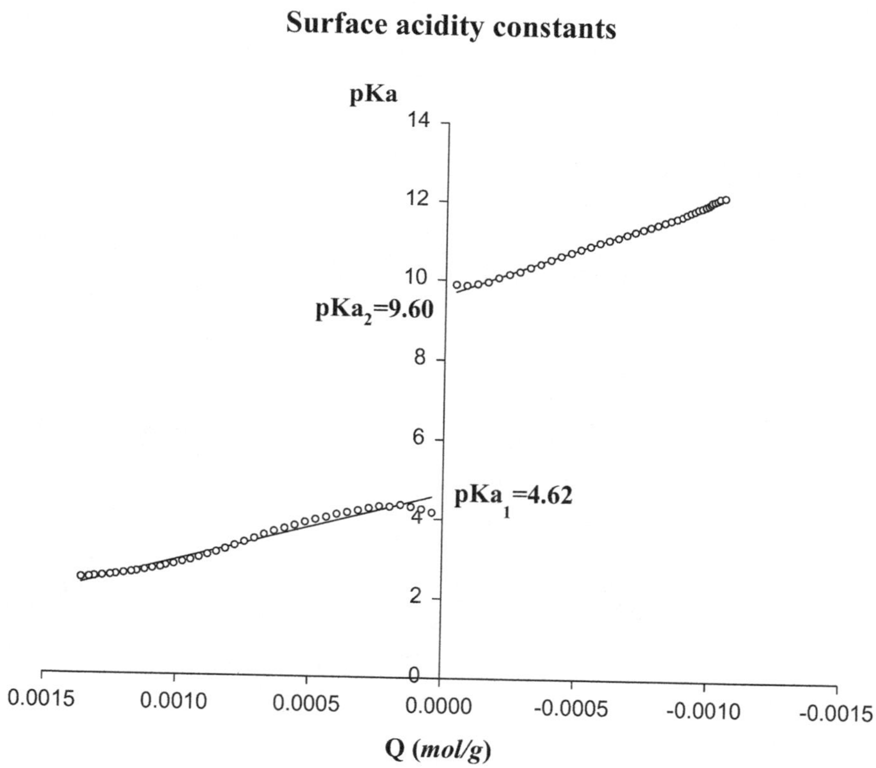


**Figure 3.6. Surface Acidity Constants for the Ferric Chloride Residual**

The buffering capacity can be determined directly from the titration curve or computed using Equation 3.19. The results of the experimental and computed β values for the ferric chloride residuals appear in Figures 3.7 and 3.8 respectively. The curves varied considerably, which indicates an error in the surface charge density and area assumed in the computation. These residuals do not have the identical physical properties of pure ferrihydrite. Similar results were observed with the alum residuals shown in Figures 3.9 to 3.13. The constants used for the alum computations were assumed to be equivalent to those of pure alumina.

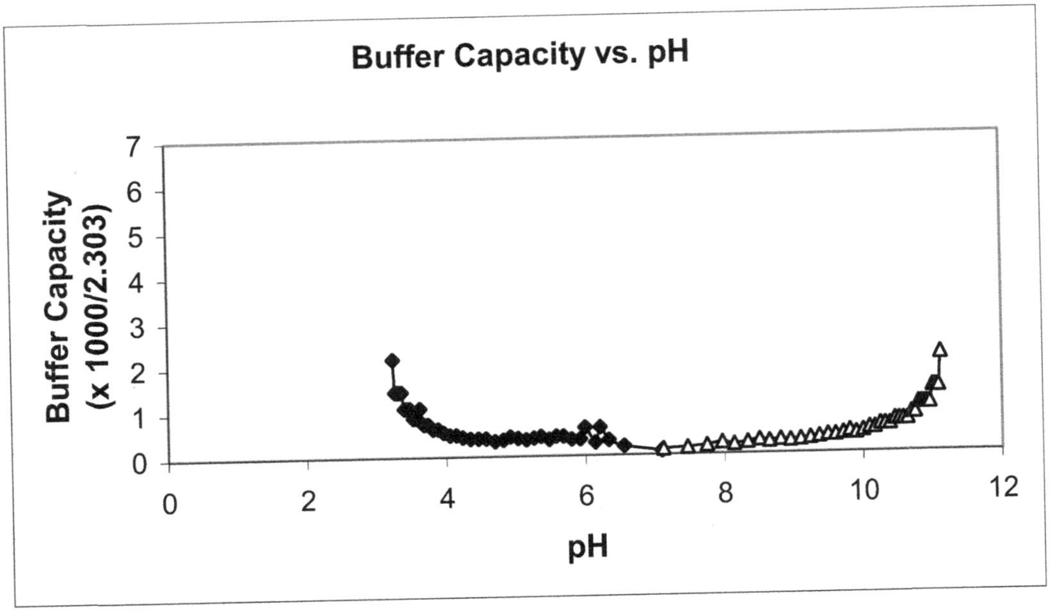


**Figure 3.7. Experiment Buffering Capacity for the Ferric Chloride Residual**

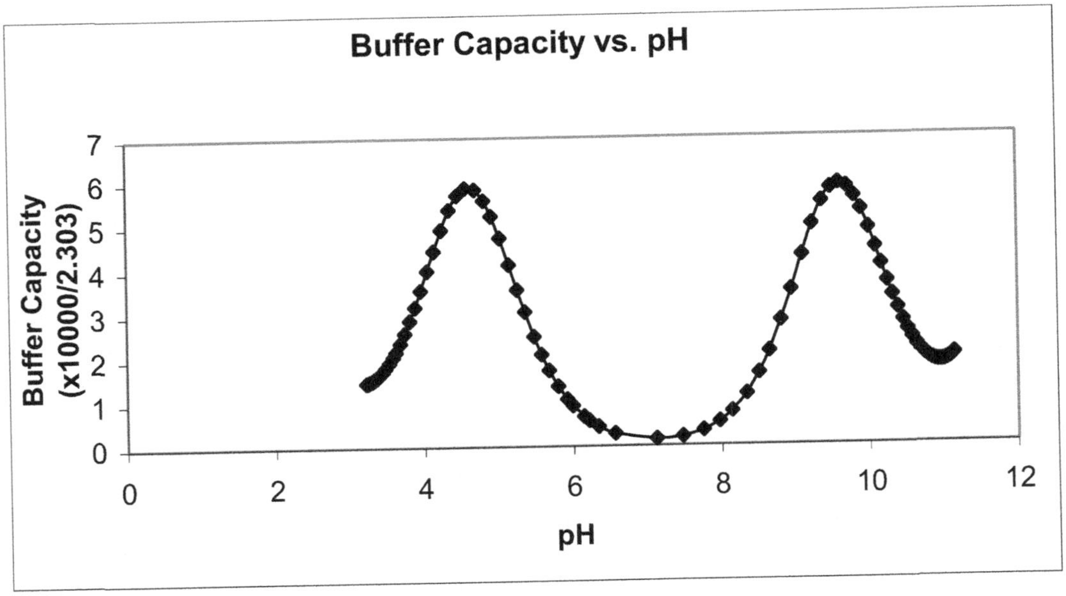


**Figure 3.8. Calculated Buffering Capacity for the Ferric Chloride Residual**

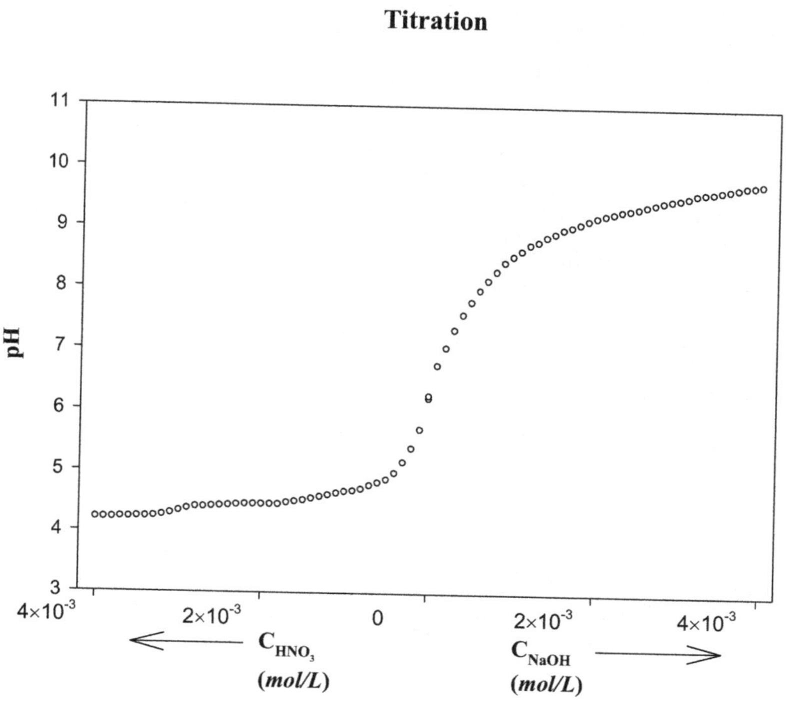


**Figure 3.9. Titration Curve for the Alum Residual**

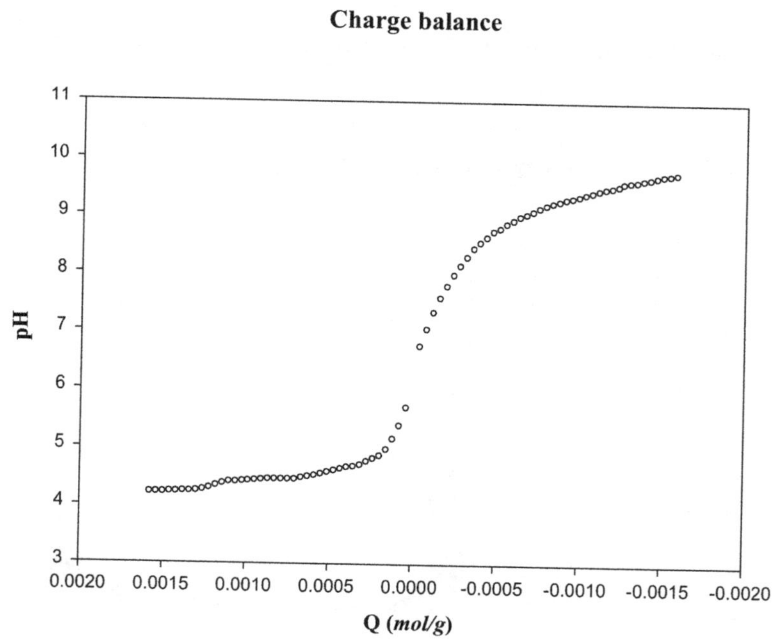


**Figure 3.10. Charge Balance for the Alum Residual**

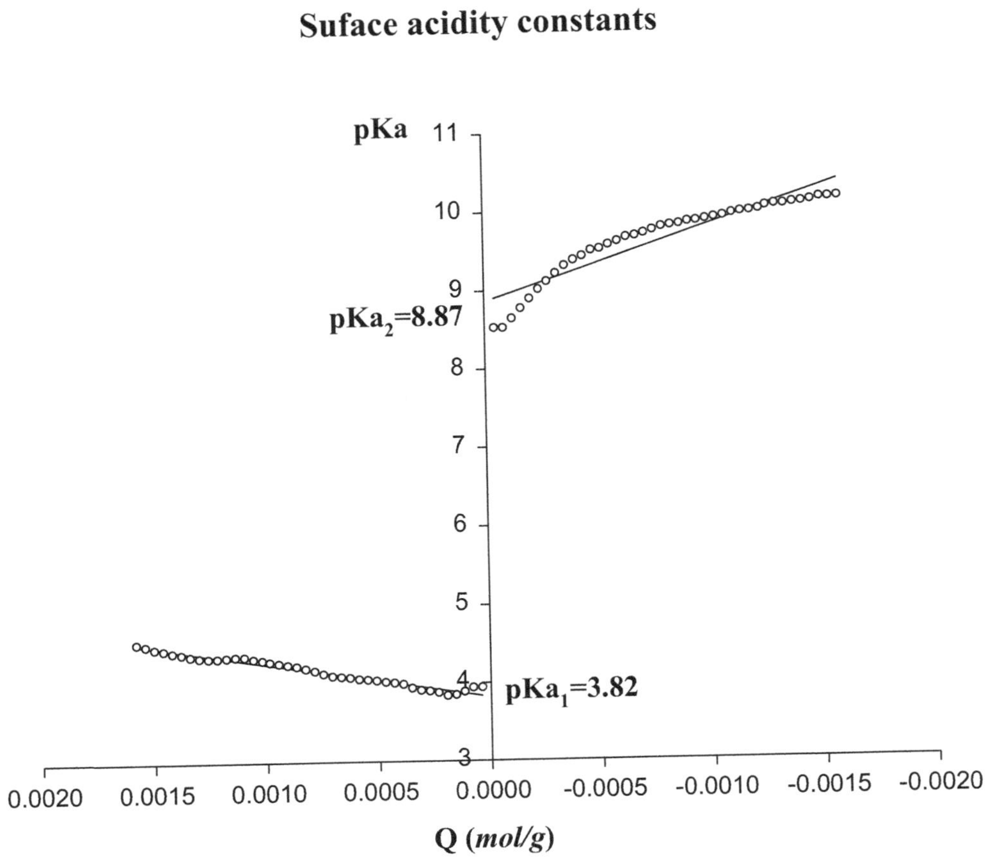


**Figure 3.11. Surface Acidity Equilibrium Constants for the Alum Residual**

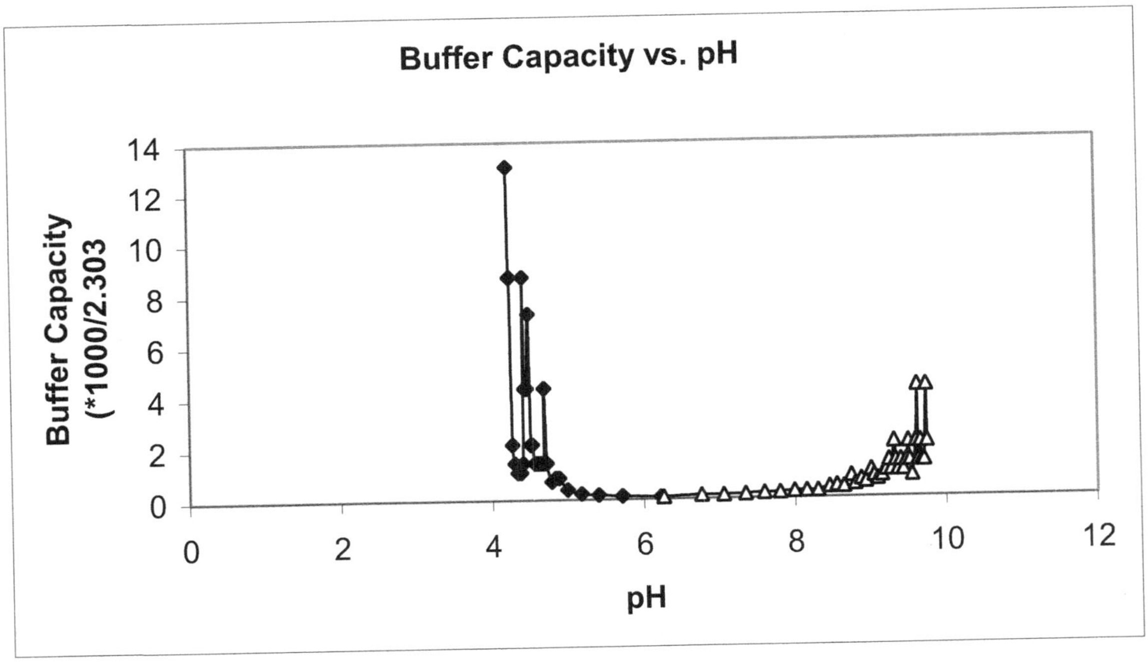


**Figure 3.12. Experimental Buffering Capacity for the Alum Residual**

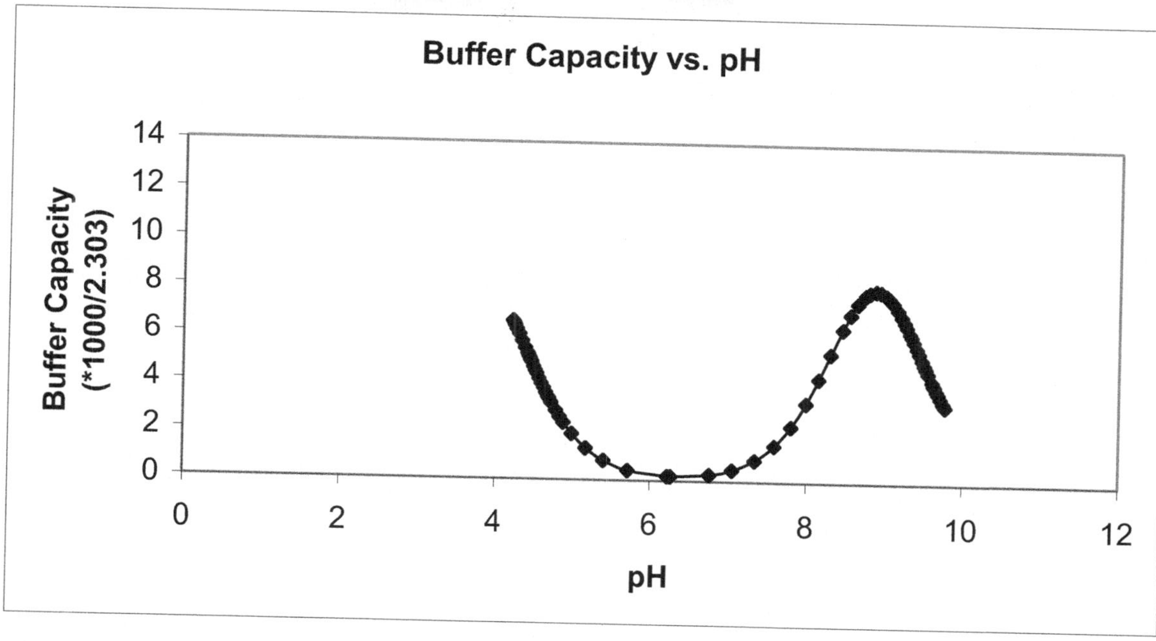


**Figure 3.13. Calculated Buffering Capacity for the Alum Residual**

## SUMMARY

Residuals that consist of aluminum or iron oxy-hydroxides can affect pH of solutions to which they are added. These effects can be predicted with a surface complex model that describes that ability of surface hydroxide species to donate or accept hydrogen ions in a similar way to acids or bases in solution. Such a model was developed for the iron and aluminum residuals. Values of $pK_{a1}$ and $pK_{a2}$ were determined to be 4.62 and 9.60, respectively, for the iron residuals and 3.82 and 8.87, respectively, for the aluminum residuals.

# CHAPTER 4: DESORPTION INHIBITION USING CALCIUM DURING COAGULATION

## INTRODUCTION

As mentioned in Chapter 2 it was hypothesized that the addition of calcium may improve (limit) desorption of arsenate from iron oxy-hydroxide based residuals. As the sorption of arsenate was best achieved using ferric chloride and the iron oxy-hydroxide system resisted desorption in most situations, the investigators chose to examine only the Fe/arsenate residual combination in this study. To explore the addition of calcium it was decided to substitute calcium hydroxide ($Ca(OH)_2$) in place of the sodium hydroxide (NaOH) used to control pH in the earlier work.

## EXPERIMENTAL METHODS

In these tests, a solution of arsenate was formulated with the initial concentration shown in Table 4.1. The precise procedure for formulating these residuals is described in Chapter 2 and will not be repeated here. The only difference between the residuals synthesized for this portion of the investigation was the use of calcium hydroxide to control pH during coagulation and co-precipitation of the arsenate. As in the methods of Chapter 2, the ferric chloride was added in a 20 to 1 mole Fe to mole As(V) ratio, while measuring the pH while mixing, and adding either NaOH or $Ca(OH)_2$ to control solution pH. Arsenate removals after sedimentation are also shown in Table 4.1. The Fe/Ca system at pH 7 did not remove arsenate as well as with the Fe/Na system although these differences may not be statistically significant.

**Table 4.1. Removal Efficiency of Arsenate Using Three Sorptive Compounds**

| Coagulants/ Chemicals | Initial As(V) conc (µg/L) | Soluble As(V) Conc after Coagulation (µg/L) | Removal Efficiency (%) |
|---|---|---|---|
| $FeCl_3 + Ca(OH)_2$ | 100,000 | 4.95 | >99.9 |
| $FeCl_3$ + NaOH | 100,000 | BDL | >99.9 |
| $Ca(OH)_2 + Na_2CO_3$ | 50,000 | 18,900 | 62.2 |

**Table 4.2. Characteristics of Each Residuals Sample**

| Sludge Type | Percent dry solids (%) | Total As (g/kg dry solids) | Total Fe (g/kg dry solids) | Total Ca (g/kg dry solids) |
|---|---|---|---|---|
| As(V) + $FeCl_3 + Ca(OH)_2$ | 17.4 | 30.9 | 500 | 24.9 |
| As(V) + $FeCl_3$ +NaOH | 11.4 | 26.7 | 531 | - |
| As(V) + $Ca(OH)_2 + Na_2CO_3$ | 60.2 | 0.793 | - | 292 |

## RESULTS

Tables 4.3 and 4.4 display the results of this study. In Table 4.3, desorption versus pH is shown. The results indicate that the Fe/Ca system was much better at retaining arsenate than the Fe/Na mixture. In the data shown in Table 4.4 the addition of calcium has eliminated the arsenate desorption in the presence of phosphate. In the phosphate study, 10 g of the residuals sample was placed in 100 mL of solution containing 1.3 mM phosphate and liquid was extracted over a 24 hour period. In Table 4.4, there was initially a high release of arsenate in the Fe/Na mixture followed by some re-adsorption. With the Fe/Ca system, a very low amount of arsenate was released, and desorption was less at higher pH than with the Fe/Na system (Table 4.3). The authors have at least two hypotheses for this behavior. One is that the calcium is forming solid phase calcium phosphate, and the other is that the calcium is altering the surface chemistry of the iron oxy-hydroxide. Another possible explanation is the formation of solid phase calcium/arsenate compounds.

**Table 4.3. Arsenate Desorption vs. pH (unit: mg As/kg dry solids)**

| Sludge Type | pH 4 | pH 6 | pH 8 | pH 10 |
|---|---|---|---|---|
| As(V) + $FeCl_3$ + $Ca(OH)_2$ | BDL | BDL | 1.46 | 45.8 |
| As(V) + $FeCl_3$ + NaOH | BDL | BDL | 2.95 | 82.5 |
| As(V) + $Ca(OH)_2$ + $Na_2CO_3$ | 373 | 39.0 | 0.16 | BDL |

**Table 4.4. Arsenate Desorption in The Presence of 1.3 mM Phosphate (unit: mg As/kg dry solids)**

| Sludge Type | Time (hr ) | | | | | | |
|---|---|---|---|---|---|---|---|
| | 0.5 | 1 | 2 | 4 | 8 | 12 | 24 |
| As(V) + $FeCl_3$ + $Ca(OH)_2$ | 0.305 | 0.269 | 0.210 | 0.150 | 0.210 | 0.507 | 0.150 |
| As(V) + $FeCl_3$ + NaOH | 221 | 123 | 51.9 | 50.2 | 39.5 | 35.9 | 30.6 |
| As(V) + $Ca(OH)_2$ + $Na_2CO_3$ | 0.514 | 0.108 | 0.108 | 0.108 | 0.108 | 0.311 | 0.108 |

## SUMMARY

Iron residuals were produced by contacting ferric iron with arsenate in the presence of calcium hydroxide or sodium hydroxide. Residuals produced in the presence of calcium retained arsenate more effectively in leaching experiments conducted over a range of pH (pH 4 – pH 10) and in the presence of 1.3 mM phosphate. The effect of calcium could be due to formation of calcium phosphate or calcium arsenate solid phases or to an interaction between calcium and the surface of the iron oxy-hydroxides that results in more effective binding of arsenic.

# CHAPTER 5: ANALYSIS OF FIELD SAMPLES

## INTRODUCTION

Water treatment residuals used for this study were supplied by three water treatment plants and two pilot plants. Most arsenic-containing residuals are generated from coagulation using alum or iron salts, adsorption (mainly using activated alumina), membrane filtration, and lime-based softening. The water treatment plants supplying residuals were El Paso Water Utilities (El Paso, TX), Public Utilities Department of the City of Billings (Billings, MT), and Helena Water Treatment Utilities (Helena, MT). The three water treatment residuals were named after their origin. Three additional arsenic-contaminated residual samples were adsorption media from two pilot plants. One of the plants was the Naval Air Station in Fallon, NV (NASF) and the other was operated by CH2M-HILL in El Paso, TX ( (CH2M-HILL - El Paso). The adsorption media used at NASF was GFH and the two media used at CH2M-HILL – El Paso, were GFO and ALCAN FS50, which is an alumina-based media. Each of the adsorption media was exhausted, which means arsenic concentration in the water leaving the adsorption column was above the MCL of 10 μg/L. The residuals from the pilot plants were named after the media type and origin.

GFH is an emerging removal technology for arsenic by adsorption in fixed bed reactors (Driehaus et al., 1998). This technique combines the advantages of the coagulation-filtration process and the fixed bed adsorption such as with activated alumina. The processing of this technology is relatively simple, and the operation can be run without the need for pH adjustment or preoxidation (Simms et al, 2000). Therefore, an unskilled person can operate the water treatment system. The process may be an acceptable treatment alternative for small-scale systems.

ALCAN FS50 is activated alumina promoted with a proprietary additive engineered to accomplish enhanced arsenic removal (Alcan Chemicals, 1998). The arsenic-adsorption capacity of ALCAN FS50 was five times greater than that of unpromoted activated alumina. Information on the water treatment facilities and residual samples is summarized in Table 5.1.

**Table 5.1. Information Summary of Water Treatment Facilities and Residual Samples**

| Residual | Type | Name of the facility |
|---|---|---|
| GFH – Fallon | GFH | Naval Air Station Fallon (NASF) – Fallon, NV |
| AA – El Paso | AA | CH2M-HILL – El Paso, TX |
| GFO – El Paso | GFO | CH2M-HILL – El Paso, TX |
| El Paso | Alum | El Paso Water Utilities – El Paso, TX |
| Billings | Ferric | Public Utilities Department – Billings, MT |
| Helena | Alum | Helena Water Treatment – Helena, MT |

## EXPERIMENTAL METHODS

### Residual Handling

Three residuals samples from water treatment plants and three residuals samples from pilot plants were utilized as target materials in this study. These samples were found to be representative of iron, alum, GFO, GFH and AA residuals. Free liquid in the residuals was removed by air-drying for the characterization studies (Kameswari et al., 2001). Each residual was placed in a clean vinyl-coated paper tray and dried under laboratory conditions for 24 hours. Subsequently, the air-dried residuals were crushed, passed through a 2-mm sieve and mixed thoroughly to produce homogeneous material. Polyethylene bottles were used to store the residuals.

## General Physical and Chemical Tests

### *Solids content*

Standard Method 2540B (APHA et al., 1998) was used to determine total solids in each residual. A portion of each residual (10 g) was dried in an oven at 105°C for 24 hours. Calculation for the percent of dry solids was conducted using equation 5.1.

$$\text{Percent of dry solids (\%)} = \frac{(C - B)}{A} \times 100 \qquad (5.1)$$

where, A = weight of raw waste (g)
B = weight of evaporating dish (g)
C = weight of dried waste + evaporating dish (g)

### *Acid digestion*

Acid digestion of sediments, sludge's, and soils, (USEPA method 3050B, USEPA, 1992) was used to establish the initial concentration of total arsenic, iron, aluminum, calcium, and manganese in each residual. One gram (dry weight) of the sample was placed in a digestion vessel on a hot plate and digested with repeated additions of nitric acid ($HNO_3$) and hydrogen peroxide ($H_2O_2$). Each digested sample was filtered using a filter paper and diluted with DI water to produce a final volume of 100 mL. Analyses were conducted for total As, Fe, Al, Ca, and Mn using flame atomic absorption spectrometry (FAAS).

### *pH measurement*

Soil and waste pH method 9045C (USEPA, 1992) was employed to determine the pH of each residual. Twenty grams (dry weight) of each residual sample was mixed with 20 mL of DI water for 20 minutes, and the pH of the suspension was measured.

## Toxicity Characteristic Leaching Procedure

TCLP (USEPA 1311, USEPA, 1992) was conducted on all of the samples using the accepted and regulated protocol (see Appendix B). Two TCLP extraction fluids were used: fluid #1 (0.1 M acetic acid, 0.064 M NaOH, pH 4.93) and fluid #2 (0.1M acetic acid, pH 2.88). Fluid #1 is usually employed for acidic wastes and fluid #2 for basic wastes. Ten grams of air-dried residual was mixed with 200 mL of leachant (liquid to solid weight ratio of 20:1) due to the small amounts of sample available. Sample and liquid mixtures in polyethylene bottles were placed into a rotary agitation apparatus for 24 hours instead of the prescribed 18 hours at room temperature (22±2°C). After the reaction, solid and liquid were separated using a 0.2 μm pore-size membrane filter. The filtrate is defined as the TCLP extract. The pH of each extract was measured and recorded. All experiments were conducted in triplicate. All extracts were kept under refrigeration (4°C) until analyzed. Arsenic concentration and speciation were analyzed by HG-AAS (see Appendix 1). Iron, aluminum, and calcium concentrations were also determined using FAAS.

## Competitive Desorption

### *Desorption with phosphate*

Based on previous work by the researchers (Chapter 2 results), phosphate tends to compete with As(V) for sorption sites on the surface of the iron (oxy)hydroxides for sorption sites, and thus tends to leach the arsenic compounds. Ten grams of air-dried residual was mixed with 200 mL of DI water, 10 mM $NaNO_3$ as an ionic strength buffer, and 0.1 M $NaH_2PO_4 \bullet H_2O$ (EM, A.C.S. reagent grade) as a phosphate source. The pH of the leachate was adjusted to 7 by the addition of $HNO_3$ or NaOH as needed. Leachate reactions subsequently were conducted for 24 hours at room temperature. All leachates were filtered through a 0.2 μm pore size membrane.

### *Desorption with sulfate*

0.1 M $Na_2SO_4$ (Fisher, A.C.S. reagent grade) as a sulfate source was mixed with 10 g of air-dried residual. The other experimental conditions were the same as those of desorption with phosphate.

### *Desorption with chloride*

These studies were conducted by the addition of 0.1 M NaCl (Fisher, A.C.S. reagent grade) to a suspension of 10 g of air-dried residual, 200 mL of DI water, and 10 mM $NaNO_3$ as an ionic strength buffer. The other experimental conditions were the same as those of desorption with phosphate.

## Effect of pH on Arsenic Leaching

Ten grams of each residual sample was added to 200 mL of DI water, with the pH adjusted by adding either NaOH or $HNO_3$ as needed to achieve the desired final pH. An ionic strength buffer, 10 mM of $NaNO_3$, was added to each mixture. A range of pH values was investigated at 4, 6, 8, and 10. Sodium acetate ($NaCH_3COO$, 10 mM) (Fisher, A.C.S. reagent grade) was employed as a pH buffer for pH 4 and 6. Sodium bicarbonate ($NaHCO_3$, 10 mM) (Fisher, A.C.S. reagent grade) was used as a pH buffer for pH 8 and 10. A mixture of the residuals and leachant was added to a polyethylene bottle and agitated by a rotary mixer for 24 hours at room temperature (22±2°C). After the reaction was completed, leachate suspensions were filtered through a 0.2 μm pore-size membrane to separate the solids and liquids. Arsenic speciation was determined using HG-AAS, and total Fe, Al, and Ca concentrations were analyzed using FAAS.

## RESULTS

### General Physical and Chemical Properties

#### *Digestion results*

The elements which can strongly influence the arsenic behavior (sorption, release, and oxidation) in residuals are iron, aluminum, calcium and manganese (Sadiq, 1997). Therefore, the concentrations of these elements initially contained in each residual sample were determined by acid digestion. The concentrations of major metals obtained from the six residuals are shown in Table 5.2.

**Table 5.2. Concentrations of Major Metals in the Residuals**

| Residual | Total As (mg/kg) | Total Fe (mg/kg) | Total Al (mg/kg) | Total Ca (mg/kg) | Total Mn (mg/kg) |
|---|---|---|---|---|---|
| GFH – Fallon | 2,680 | 521,000 | 212 | 1,510 | 741 |
| AA – El Paso | 33.2 | 20,600 | 298,000 | 5,640 | 10.5 |
| GFO – El Paso | 2,290 | 503,000 | 993 | 5,690 | 852 |
| El Paso | 15.3 | 24,000 | 18,100 | 35,800 | 425 |
| Billings | 148 | 50,000 | 30,300 | 18,400 | 713 |
| Helena | 402 | 20,300 | 45,800 | 1,860 | 1,350 |

As expected, the highly elevated concentrations of total arsenic and total iron in GFH and GFO of the pilot plant studies were detected due to the high arsenic adsorption capacity and the adsorption media being made principally from ferric hydroxide. From Table 4.2, the contents of arsenic in the GFH and GFO were greater than that in the AA on the basis of weight. Chemicals added typically vary in water treatment facilities. Therefore, it is hard to find solid relationships between the arsenic removal, type of media and digestion results.

#### *Results of pH and solid content*

The pH and solid content of each air-dried residual are shown in Table 5.3. Water treatment residuals are typically dewatered prior to disposal using either air-drying or mechanical dewatering (Hsieh and Raghu, 1997). The most common processes for dewatering are sand drying beds, lagoons, belt filter presses, centrifuges, pressure filter presses, and vacuum.

**Table 5.3. pH and Solid Content of the Residuals**

| Residual | pH | Solid content (%) |
|---|---|---|
| GFH – Fallon | 7.27 | 69.1 |
| AA – El Paso | 7.55 | 83.6 |
| GFO – El Paso | 7.85 | 67.9 |
| El Paso | 7.85 | 96.5 |
| Billings | 7.44 | 91.9 |
| Helena | 7.93 | 80.9 |

Knowledge of pH is critical in predicting arsenic behavior in a treatment residual, as arsenic adsorption and release reactions are very pH dependent. The pH values of the residual samples were measured according to the USEPA method 9045C (USEPA, 1992). The pH values of all of the residuals were found to be around neutral (between 7 and 8). This result was expected due to the previously mentioned buffering ability of metal oxy-hydroxide residuals.

**TCLP Analysis**

The field residuals were subjected to two types of leaching solutions for the toxicity characteristics leaching procedure (TCLP). TCLP extraction solution #1 (pH=4.93) was developed for acidic solid wastes and TCLP extraction solution #2 (pH=2.88) is for basic wastes. However, in the current study TCLP #1 and TCLP #2 were used for all residuals as pH values for the residuals were around neutral. Results of TCLP #1 and TCLP #2 are presented in Table 5.4 and Table 5.5, respectively. All testing was conducted in triplicates. Total arsenic recoveries were calculated using total arsenic concentrations presented in Table 5.2.

**Table 5.4. Results of TCLP Extraction Solution #1**

| Residual | Final pH | As(III) (mg/kg) | As(V) (mg/kg) | Total As (%) | Fe (mg/kg) | Al (mg/kg) | Ca (mg/kg) |
|---|---|---|---|---|---|---|---|
| GFH – Fallon | 5.44 | < 0.02 | 0.37 | 0.01% | < 1 | < 20 | 1,070 |
| AA – El Paso | 5.18 | 0.10 | 0.07 | 0.51% | 0.60 | 157 | 5,340 |
| GFO – El Paso | 5.14 | 0.29 | 0.15 | 0.02% | < 1 | < 20 | 8,410 |
| El Paso | 6.61 | 0.06 | 0.24 | 1.96% | < 1 | < 20 | 11,100 |
| Billings | 5.94 | 0.31 | 0.79 | 0.74% | 3.12 | < 20 | 8,530 |
| Helena | 5.48 | 2.49 | 6.13 | 2.14% | 47.8 | 151 | 817 |

As(III) and As(V) concentrations in the extracted solutions were both higher with TCLP #2 than with TCLP #1, except for the GFO – El Paso residual in which leached As(V) concentration was higher with TCLP #1. As(III) was the major As species of the total soluble As in AA – El Paso and GFO – El Paso residuals, even though the redox status of the mixed TCLP extract and residuals was assumed to be oxidizing. It is thought that the main reason why these results were obtained is that AA and GFO media were utilized during pilot plant studies in El Paso for arsenic removal from groundwater and thus, it is assumed they contained much larger portions of As(III). However, As(V) was the dominant As species released from the other residuals samples. When a higher As concentration was released from a residual, a higher concentration of a solubilized adsorbent, such as Fe or Al hydroxide, was also observed. Therefore, it is concluded that As desorption behavior is controlled by the Fe or Al oxy-hydroxides. Soluble Ca concentration was high with all residuals from both TCLP tests, due to the high solubility of $Ca(OH)_2$ and $CaCO_3$ at low pH. Mn concentrations of the residuals were lower than other elements (ref. Table 5.2), and Mn concentrations in the leachates were very low. Thus, measurement of the Mn concentration was not conducted in further tests. The final pH of each extract was lower with TCLP #2 than with TCLP #1 as expected.

**Table 5.5. Results of TCLP Extraction Solution #2**

| Residual | Final pH | As(III) (mg/kg) | As(V) (mg/kg) | Total As (%) | Fe (mg/kg) | Al (mg/kg) | Ca (mg/kg) |
|---|---|---|---|---|---|---|---|
| GFH – Fallon | 3.50 | < 0.02 | 0.62 | 0.02% | 132 | < 20 | 1,450 |
| AA – El Paso | 4.01 | 0.16 | < 0.02 | 0.48% | 6.38 | 3,610 | 3,520 |
| GFO – El Paso | 3.76 | 0.33 | < 0.02 | 0.01% | 2.83 | < 20 | 5,820 |
| El Paso | 4.93 | 0.07 | 0.20 | 1.76% | 286 | < 20 | 28,500 |
| Billings | 4.30 | 0.72 | 1.74 | 1.66% | 3,410 | 203 | 12,300 |
| Helena | 3.97 | 4.97 | 14.2 | 4.77% | 919 | 4,510 | 877 |

From these data, it is concluded that pH is critical for the release of As, Fe, Al, and Ca. In oxidized systems, the solubility, adsorption and movement of arsenic (As(III) and As(V) both) are controlled predominantly by adsorption reactions with oxide minerals such as Fe and Al oxides (Loeppert et al., 2003; Sadiq, 1997; Masscheleyn et al., 1991; Artiola et al., 1990). At low pH (3-5), solubility of the Fe and Al oxides increases, and thus the concentration of inorganic arsenic released in the leachates is expected to also increase. However, maximum As(V) adsorption on Fe and Al hydroxides is usually achieved at low pH (4-5) (Raven et al., 1998; Anderson et al., 1976). This means that the conditions of the TCLP may be such to minimize arsenic release from iron residuals.

Calcium may precipitate arsenic in the form of calcium arsenate at high pH. However, as shown in Tables 5.4 and 5.5, released calcium concentrations were significant due to the low pH of the leachates. Calcium carbonates are soluble and unstable at low pH conditions and consequently, they play a minor role in arsenic solubility from acidic residuals (Sadiq, 1997).

## Competitive Desorption Analysis

### *Competitive desorption with phosphate*

The mobilization of bound arsenic on adsorbents such as Fe and Al oxy-hydroxides is strongly influenced by the presence of ligands that can compete with arsenic for surface sorption sites. In particular, phosphate tends to compete with As(V) for sorption sites on the surfaces of the metal oxy-hydroxides (equation 5.2) and thus, tends to extract the As(V) compounds (Loeppert et al, 2003; Hongshao and Stanforth, 2001; Hiemstra and Riemsdijk, 1999; Hingston et al., 1971).

$$\text{Fe-oxide-AsO}_4\text{H} + \text{HPO}_4^{2-} \rightarrow \text{Fe-oxide-PO}_4\text{H} + \text{HAsO}_4^{2-} \qquad (5.2)$$

The sorption behavior of phosphate and As(III) are different, and a significantly higher concentration of bound As(III) compared to As(V) was observed on iron minerals in the presence of 0.1 M sodium phosphate (Jackson and Miller, 2000; Jain and Loeppert, 2000). These studies indicated that there might be specific sites for As(III) on the iron oxide minerals.

Figure 5.1(a) presents the soluble concentrations of As(III) and As(V) and Figure 5.1(b) presents the soluble concentrations of Fe, Al, and Ca released by the competition with phosphate. Extraction solutions were adjusted to pH 7, and the final pH values after 24 hours of reaction were between 7.2 and 7.6. These results indicated that As(V) was the dominant arsenic species extracted by phosphate. GFH – Fallon and GFH – El Paso residuals contained high concentrations of total arsenic (ref. Table 5.2) and thus, the concentration of As(V) released from the two residuals was much higher than that from the other residuals. Arsenic concentrations extracted by phosphate from Fe oxide minerals usually increase at low and high pH conditions (Loeppert et al., 2003). Therefore, it is expected that higher As concentration would be released at low and high pH than the leached As concentration at approximately neutral pH in this study.

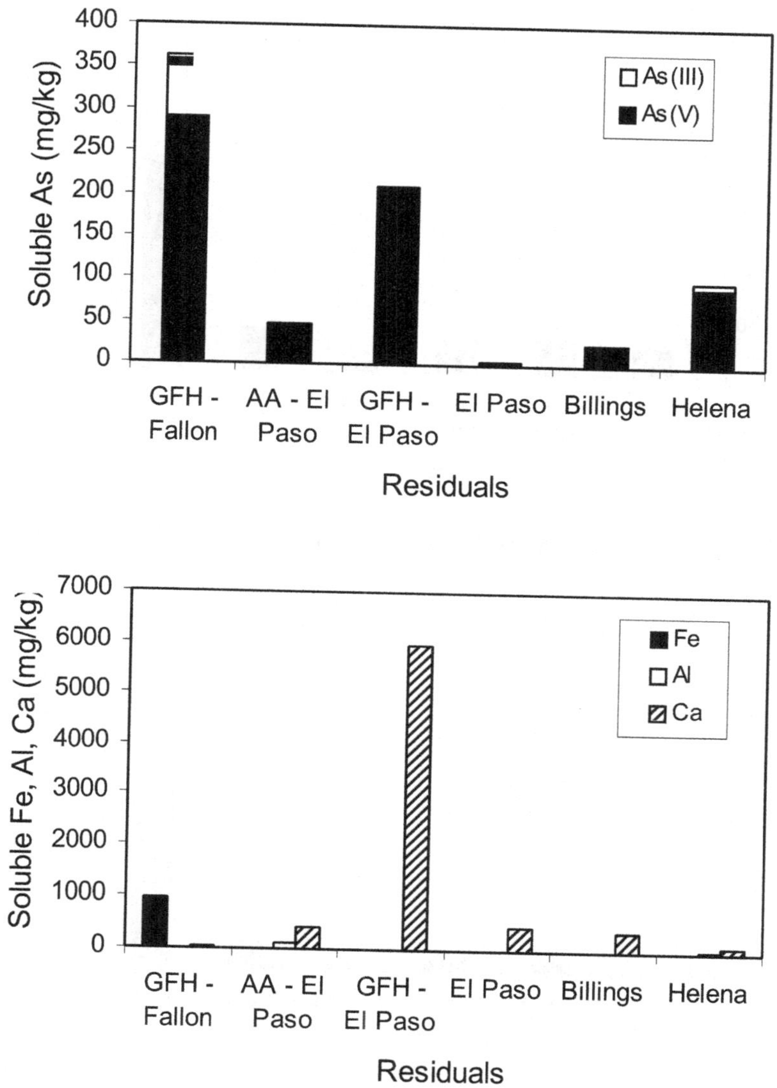


**Figure 5.1. Competitive Desorption with Phosphate: (A) Soluble As Concentration, and (B) Soluble Fe, Al, and Ca Concentrations**

***Competitive desorption with sulfate and chloride***

The results of arsenic leaching in the presence of sulfate and chloride are shown in Figure 5.2 and Figure 5.3. Any supporting data in which the presence of sulfate ions can influence the release of As were not found. However, it was reported that As(III) and As(V) adsorption was decreased in the presence of sulfate and that this trend was more obvious at lower pH (Wilkie and Hering, 1996). It is concluded that the effect of sulfate on As desorption is minor even though it can influence As sorption adversely. An enhanced release of arsenic in the presence of chloride was also not observed (Figure 5.3). The amount of As released in the presence of sulfate and chloride was much lower than that in the presence of phosphate in comparing Figures 5.1 to 5.2 and Figures 5.1 to 5.3.

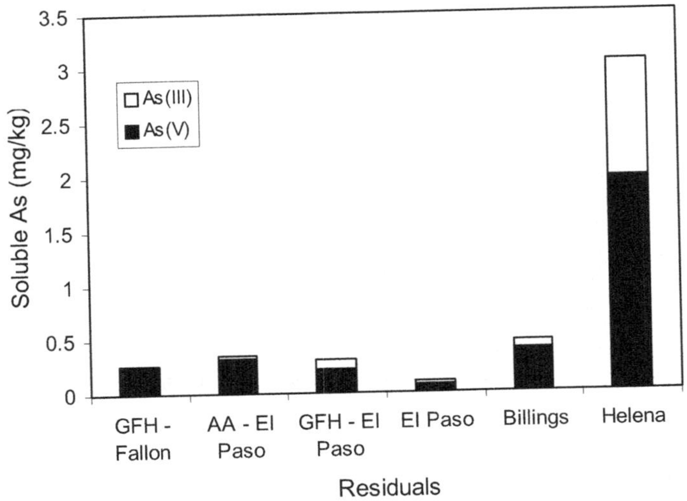


**Figure 5.2. Competitive Desorption with Sulfate: Soluble As Concentrations**

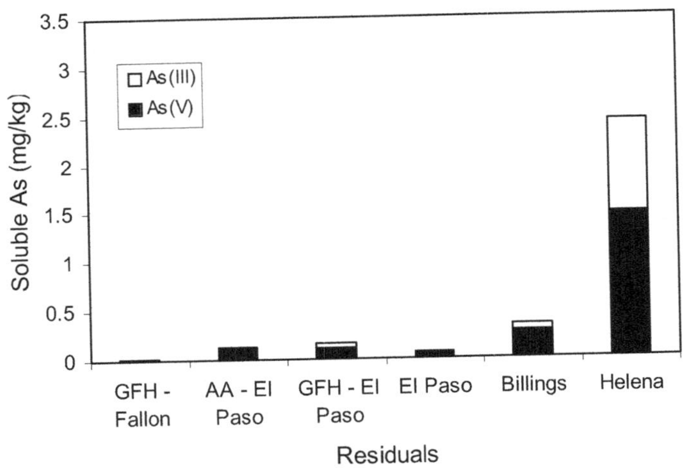


**Figure 5.3. Competitive Desorption with Chloride: Soluble As Concentrations**

**Effect of pH on Arsenic Leaching Analysis**

Figure 5.4 shows the soluble As(III) and As(V) concentrations, and Tables 5.6, 5.7, and 5.8 present the influence of pH on leached Fe, Al, and Ca concentrations, respectively. Soluble arsenic and the other major metals (Fe, Al, and Ca) concentrations depended on the pH of the system.

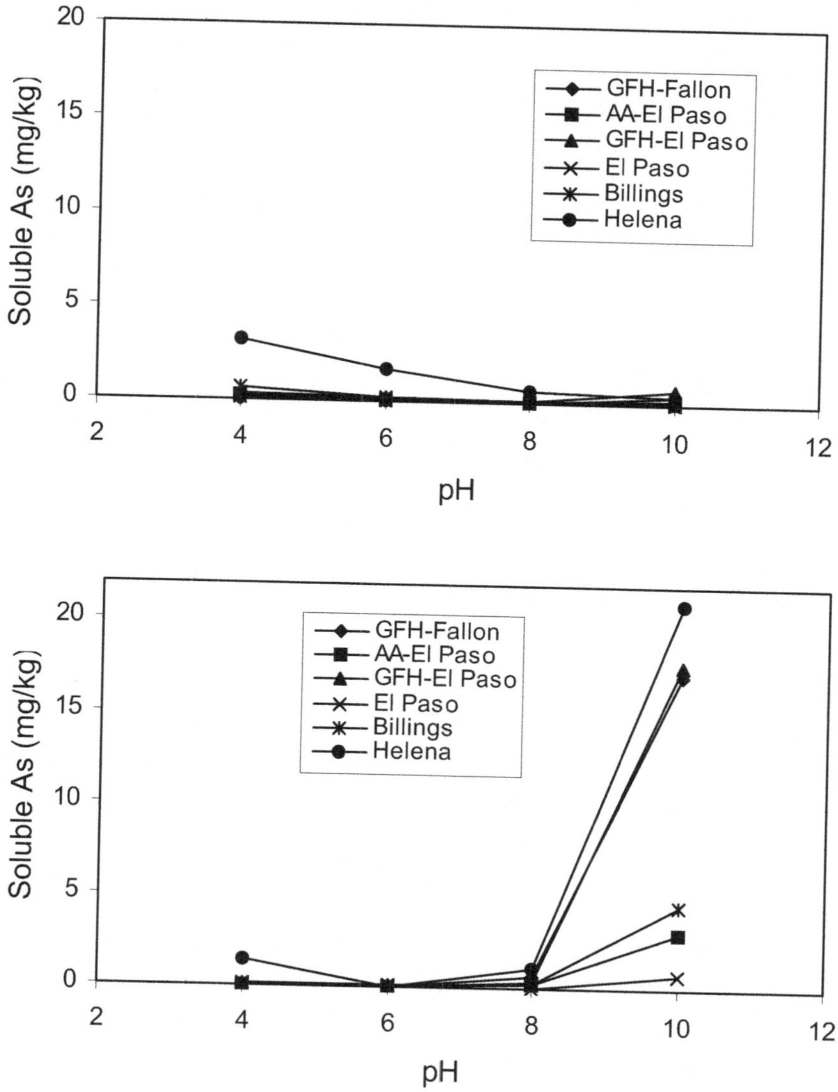


**Figure 5.4. Effect of pH on Arsenic Leaching: (A) As(III), and (B) As(V)**

**Table 5.6. Effect of pH on Leaching: Fe**

| pH | GFH – Fallon (mg/kg) | AA – El Paso (mg/kg) | GFO – El Paso (mg/kg) | El Paso (mg/kg) | Billings (mg/kg) | Helena (mg/kg) |
|---|---|---|---|---|---|---|
| 4 | < 1 | < 1 | < 1 | 1,570 | 2,260 | 504 |
| 6 | < 1 | < 1 | < 1 | < 1 | < 1 | 4 |
| 8 | < 1 | < 1 | < 1 | < 1 | < 1 | < 1 |
| 10 | 457 | < 1 | 1,360 | < 1 | < 1 | 12 |

**Table 5.7. Effect of pH on Leaching: Al**

| pH | GFH – Fallon (mg/kg) | AA – El Paso (mg/kg) | GFO – El Paso (mg/kg) | El Paso (mg/kg) | Billings (mg/kg) | Helena (mg/kg) |
|---|---|---|---|---|---|---|
| 4 | < 20 | 1,140 | < 20 | 85 | 304 | 1,670 |
| 6 | < 20 | < 20 | < 20 | < 20 | < 20 | < 20 |
| 8 | < 20 | < 20 | < 20 | < 20 | < 20 | < 20 |
| 10 | < 20 | 72 | < 20 | < 20 | 36 | 658 |

**Table 5.8. Effect of pH on Leaching: Ca**

| pH | GFH – Fallon (mg/kg) | AA – El Paso (mg/kg) | GFO – El Paso (mg/kg) | El Paso (mg/kg) | Billings (mg/kg) | Helena (mg/kg) |
|---|---|---|---|---|---|---|
| 4 | 1,640 | 7,430 | 9,420 | 31,500 | 13,300 | 1,050 |
| 6 | 684 | 4,300 | 6,760 | 25,300 | 4,920 | 510 |
| 8 | 16 | 959 | 944 | 432 | 755 | 183 |
| 10 | 10 | 108 | 179 | 133 | 112 | 100 |

Higher arsenic concentrations were observed at low and high pH (4 and 10). As(III) concentrations were highest at pH 4, and soluble As(V) concentrations were highest at pH 10. The maximum adsorption for As(V) and As(III) typically occurs at around pH 5 and around 9, respectively. Results of soluble Fe and Al concentrations were similar to those of arsenic. Interestingly, higher leached Fe and As(V) concentrations in GFH – Fallon and GFO – El Paso residuals were observed simultaneously at pH 10 because dissolution of Fe hydroxides occurred and, subsequently, As(V) was released. Higher soluble Al concentrations were observed at pH 4 than at pH 10. Leached calcium concentrations were substantially increased as pH decreased due to the increase of calcium solubility. The results obtained in this study were very similar to those of other investigators (Carbonell-Barrachina et al., 1999; Masscheleyn et al., 1991)

The As(III) and As(V) desorption trends can be explained by the dissolution of Fe and Al hydroxides (i.e., increase of solubility) and the charge of arsenic species. The solubility diagram of amorphous Fe hydroxide as a function of pH is shown in Figure 5.5, which was constructed based on the hydrolysis constants in Table 5.9. Activity coefficients for the species were ignored, so activities and concentrations were assumed to be equal.

In Table 5.9, $^*\beta$ is the hydrolysis constant. For the general reaction,

$$M + iHL = ML_i + iH^+ \quad (5.3)$$

where, M = metal, and L = ligand.

$$^*\beta_i = \frac{[ML_i][H^+]^i}{[M][HL]^i} \quad (5.4)$$

**Table 5.9. The Hydrolysis Constants for the Fe Hydroxides (Langmuir, 1997b)**

| Reaction | $-\log{}^*\beta_i$ |
|---|---|
| $Fe^{3+} + H_2O = FeOH^{2+} + H^+$ | 2.19 |
| $Fe^{3+} + 2H_2O = Fe(OH)_2^+ + 2H^+$ | 5.67 |
| $Fe^{3+} + 3H_2O = Fe(OH)_3^0 + 3H^+$ | 12.56 |
| $Fe^{3+} + 4H_2O = Fe(OH)_4^- + 4H^+$ | 21.6 |

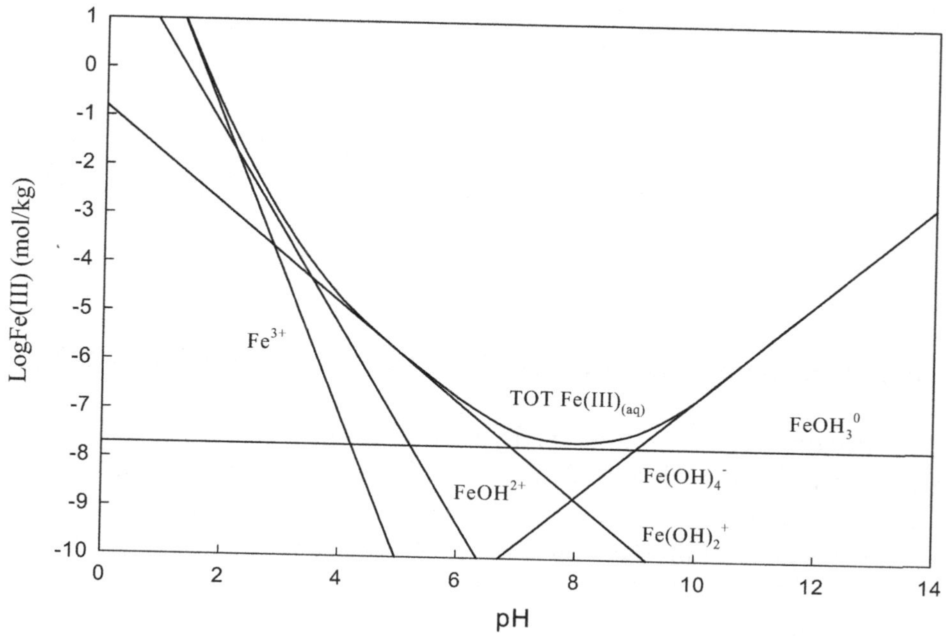


**Figure 5.5. Solubility of Amorphous $Fe(OH)_3$, $K_{sp} = 10^{-37.1}$, as a Function of pH at 25°C**

The solubility of Al hydroxide as a function of pH (Figure 5.6) was drawn using the same method used for the solubility of Fe hydroxide. Hydrolysis constants are summarized in Table 5.10. The phenomenon of the increase in solubility of the solids, such as Fe and Al hydroxides, at low and high pHs is termed amphoteric behavior, where solids are dissolved to form cationic species at low pH and form anionic species at high pH (Langmuir, 1997b).

**Table 5.10. The Hydrolysis Constants for the Al Hydroxides (Langmuir, 1997b)**

| Reaction | $-\log{}^*\beta_i$ |
|---|---|
| $Al^{3+} + H_2O = AlOH^{2+} + H^+$ | 5 |
| $Al^{3+} + 2H_2O = Al(OH)_2^+ + 2H^+$ | 10.1 |
| $Al^{3+} + 3H_2O = Al(OH)_3^0 + 3H^+$ | 16.9 |
| $Al^{3+} + 4H_2O = Al(OH)_4^- + 4H^+$ | 22.7 |

An equation for the $H^+$ enhanced dissolution reaction can be expressed as in equation 5.5 (Loeppert et al., 2003). This equation describes the formation of cationic species by the dissolution of metal oxy-hydroxides.

$$\text{M-oxide-AsO}_4 + H^+ \rightarrow M^{3+} + H_2O + As_{aq} \qquad (5.5)$$

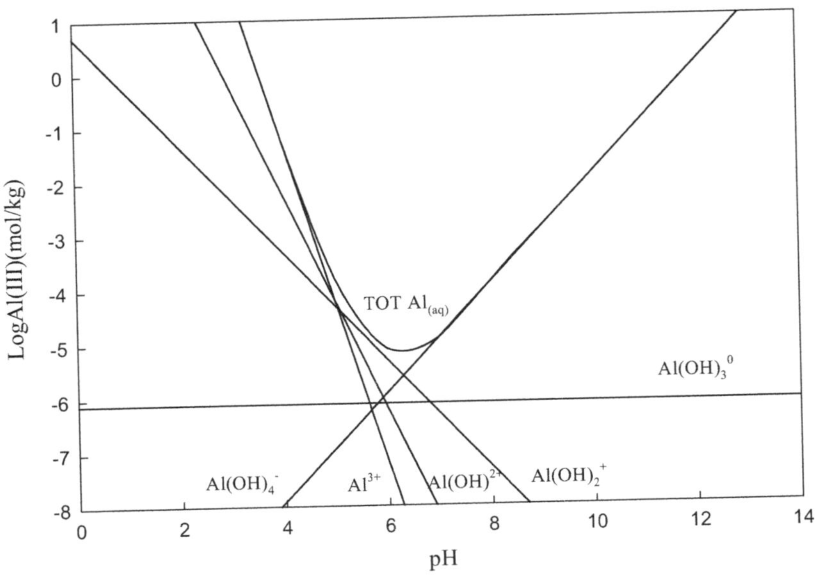


**Figure 5.6. Solubility of Amorphous $Al(OH)_3$, $K_{sp} = 10^{-31.2}$, as a Function of pH at 25°C**

During the dissolution of Fe and Al hydroxides, surface arsenic species can also be dissolved. At pH 4, the predominant As(III) species is $H_3AsO_3^0$, which has a neutral charge. The surfaces of Fe and Al oxy-hydroxides are positively charged at low pH and the charge of the dominant As(III) species is neutral. Therefore, with the dissolution of metal hydroxides at pH 4, the positively charged surfaces of metal oxy-hydroxides and the neutrally charged dominant As(III) species facilitate the release of bound arsenic, and readsorption of leached As(III) cannot readily occur. At pH 4, the prevalent species for As(V) is $H_2AsO_4^-$, which can be easily adsorbed onto the metal hydroxides. Therefore, soluble As(V) concentration in the leachate was lower than that of As(III) at low pH (pH 4).

At high pH, anionic species of Fe and Al oxy-hydroxides are increased by the $OH^-$ enhanced dissolution process, which can be expressed as in equation 5.6 (Loeppert et al., 2003; Jackson and Miller, 2000).

$$\text{M-oxide-}AsO_4 + OH^- \rightarrow \text{M-}(OH)_4^- + As_{aq} \quad (5.6)$$

The surfaces of Fe and Al hydroxides are negatively charged at high pH. Therefore, the increasingly negative surface potential with increasing pH produces unfavorable conditions for the adsorption of anionic arsenic species such as $HAsO_4^{-2}$ for As(V) and $H_2AsO_3^-$ for As(III). Under these conditions, arsenic ions are readily leached from the sorption sites. In brief, the increasingly negative surface charge of the metal oxy-hydroxides with increasing pH promotes the desorption of As(V). The neutrally and negatively charged species of As(III) are predominant at pH 10, and thus As(III) should be released due to repulsion from the negatively charged surface of the metal oxy-hydroxides. At pH 10, the lower concentration of leached As(III) compared to As(V) was attributable to the lower concentrations of As(III) in the residuals and the presence of neutrally charged species of As(III).

## SUMMARY

Residuals from six water treatment plants were examined and they included GFH, GFO, AA, aluminum and iron residuals. They contained both arsenate and arsenite at total concentrations from 15 to 2,680 mg/kg. Experiments with both TCLP leaching solutions showed higher amounts of arsenic released with solution 2, probably due to its lower pH. Generally more As(V) was released than As(III) was released, but the distribution of arsenic species in the unleached residuals was not known.

Phosphate at 0.1 M concentration increased the amount of arsenate released, but sulfate and chloride at the same concentration had little effect.

Arsenite release was greatest at pH 4 and arsenate release was greatest at pH 10. Release of iron at pH 10 indicated degradation of GFH and GFO. At neutral pH (pH 6 and 8), the soluble concentrations of As, Fe and Al were very low due to the insolubility of the Fe and Al hydroxides and the predominant arsenic species. When the disposal of the arsenic-tainted residuals is planned without any post-treatment such as solidification/stabilization, a neutral pH should be maintained to minimize arsenic solubility and mobilization.

# CHAPTER 6: STABILIZATION TECHNIQUES FOR IMMOBILIZATION OF ARSENIC

## INTRODUCTION

The goals of this portion of the study were to analyze a variety of water treatment residuals containing arsenic and to develop suitable methods for immobilizing arsenic in them. Two types of residuals were investigated (GFO, AA) and three stabilization methods were studied (lime, OPC, lime and OPC). The extent of immobilization was measured by two extraction procedures (TCLP, phosphate solution).

## EXPERIMENTAL METHODS

### Preparation of Residuals

The residuals obtained for the study were maintained in a refrigerated environment. Free liquid in the residuals was removed via air-drying for the solidification and stabilization (S/S) experiments (Kameswari et al., 2001). Air-dried residuals were then crushed, passed through a 2 mm sieve and thoroughly mixed to produce homogeneous samples. All S/S experiments were carried out on the basis of 10 g of air-dried residual. Added-water to residuals ratio varied from 0.5 to 1, depending on the amount of binder used. Residuals and binder(s) were mixed thoroughly by hand with an acid-washed glass bar before addition of the water. The residuals-binder mixture was placed in an acid-washed plastic cup and cured at room temperature for 7 days. After curing, all solidified samples were crushed to a particle size < 2 mm (using a 2-mm nominal pore-size sieve) and stored in polyethylene bottles separately until subjected to extraction tests. All experiments were performed in triplicate.

### Addition of Lime Only

Quicklime (CaO, calcium oxide) is mainly employed for S/S in field-scale applications due to economic factors. Mixing CaO and water produces a slurry of $Ca(OH)_2$ in a process termed slaking, which is an exothermic reaction.

$$CaO + H_2O = Ca(OH)_2 + \text{heat} \qquad (6.1)$$

Care must be taken when CaO is handled since high temperatures are generated (Davis and Cornwell, 1998). Due to this possible safety problem, hydrated lime ($Ca(OH)_2$) was used in this study. Similar results were obtained when $Ca(OH)_2$ was used instead of CaO for S/S of arsenic in waste (Dutre and Vandecasteele, 1995). In the current study, $Ca(OH)_2$ (EM, A.C.S. reagent grade) was investigated by adding increasing amounts (1 to 10g per 10g of air-dried residual) to the residuals. After curing for 7 days, extraction tests were conducted to determine the optimum amount of lime for the control of arsenic leaching.

### Addition of OPC Only

Ordinary Portland cement (OPC) is usually used as the primary binder in cement-based stabilization. The OPC used in this study was obtained from the Quikrete Company. Different ratios of cement to residual (weight basis) were used: 0.1, 0.2, 0.3, and 0.5, and the optimum amount of cement was determined.

### Addition of Lime and OPC

Based on the results of the stepped additions of lime and OPC only, the optimum amounts of lime and OPC were selected and added to each residual sample.

### Extraction Tests

To examine the effectiveness of the S/S of each residual sample, extraction tests were performed. Untreated residuals and all solidified solids were subjected to two extraction tests: TCLP #2 (0.1 M acetic acid, pH = 2.88) extraction solution, and the worst-case arsenic leaching condition (0.1 M $NaH_2PO_4$, pH = 10) as discussed in Chapter 2. The liquid to solid weight ratio was 20, i.e., 200 mL of leachant and 10 g of air-dried residual (weight of binder was ignored for the comparisons of leaching of raw residual to that of solidified solid). All solidified specimens were pulverized to the particle size < 2 mm. The extraction period was 24 hours for all extraction tests. Leaching reactions were conducted in polyethylene bottles by a rotary shaker. After extraction, the leachate suspensions were filtered with a 0.2 μm membrane filter to separate liquid and solids. Arsenic speciation and Fe, Al, and Ca concentrations were determined according to the analytical methods described in the Appendices and previous chapters. The variables used in the extraction tests and subsequent analyses conducted are summarized in Table 6.1.

**Table 6.1. Extraction Tests Examined and Subsequent Analyses Conducted**

| Parameter | Description |
|---|---|
| Types of leachant | (i) TCLP extraction #2 (0.1 M acetic acid, pH = 2.88)<br>(ii) Phosphate extraction (0.1 M $NaH_2PO_4$, pH = 10) |
| Leachant-to-residual ratio | 20:1 (200 mL: 10 g) |
| Surface area of residual | Particle size smaller than 2.00 mm |
| Extraction period | 24 hours |
| Extraction vessel | Polyethylene bottle |
| Temperature | Room temperature (22±3°C) |
| Separation of liquid and solids | 0.2-μm pore size membrane filter with syringe |
| Storage of samples | In a refrigerator (4°C) |
| Arsenic speciation | Using HG-AAS within 2 days after collecting samples |
| Fe, Al, Ca analyses | Using FAAS within 3 days after collecting samples |

## RESULTS OF THE STABILIZATION TECHNIQUES

The residuals employed for S/S treatment were GFH – Fallon and AA – El Paso. The selection criteria were arsenic concentration and the amount of available residual. In this section, GFH – Fallon was noted as GFH and AA – El Paso as AA. Lime (hydrated lime, $Ca(OH)_2$) and ordinary Portland cement (OPC) were utilized as binders for S/S. TCLP extraction solution #2 (0.1 M acetic acid, pH = 2.88), noted as TCLP #2, and phosphate solution (0.1 M $NaH_2PO_4$, pH=10), noted as 0.1M phosphate, were used for the extraction studies.

**Addition of Lime Only**

***Granular Ferric Hydroxide (GFH)***

In the first step for optimization of the S/S procedure, $Ca(OH)_2$ was added to the residual in 1 g increments per 10 g of air-dried residual, and the solidified materials were cured for 7 days under laboratory conditions. The effects of lime addition on arsenic leaching of GFH are shown in Table 6.2 and Table 6.3.

**Table 6.2. Extracted As, Fe, and Ca Concentrations and Final pH by TCLP #2 Depending on Lime Addition: GFH**

| Lime addition (g) | As(III) (μg/L) | As(V) (μg/L) | Total As (%) | Fe (mg/L) | Ca (mg/L) | Final pH |
|---|---|---|---|---|---|---|
| 0 | < 1 | 23 | 0.017% | 2 | 56 | 3.53 |
| 1 | < 1 | 9 | 0.007% | < 0.05 | 1,620 | 6.30 |
| 2 | < 1 | 4 | 0.003% | < 0.05 | 1,570 | 8.36 |
| 3 | < 1 | 3 | 0.002% | < 0.05 | 1,570 | 11.08 |
| 4 | < 1 | 3 | 0.002% | < 0.05 | 1,920 | 11.09 |
| 5 | < 1 | 2 | 0.001% | < 0.05 | 1,820 | 11.22 |
| 6 | < 1 | 2 | 0.001% | < 0.05 | 2,010 | 11.28 |
| 7 | < 1 | 1 | 0.001% | < 0.05 | 1,970 | 11.32 |
| 8 | < 1 | 2 | 0.001% | < 0.05 | 1,880 | 11.35 |
| 9 | < 1 | 2 | 0.001% | < 0.05 | 1,990 | 11.37 |
| 10 | < 1 | 1 | 0.001% | < 0.05 | 2,080 | 11.38 |

As shown in Table 6.2, leached arsenic concentrations were very low by the TCLP #2 (pH=2.88) extraction test. Therefore, it was almost impossible to find any effect of lime addition on arsenic leaching. However, the effect of lime addition was obvious when solidified materials were subjected to 0.1 M phosphate (pH=10) extraction solution. The final pH values of the leachates by phosphate extraction were in the range of 10.2 to 13.2, depending on the lime treatment. The total soluble As and Ca concentrations in the phosphate extractions as a function of lime addition are shown in Figure 6.1.

**Table 6.3. Extracted As, Fe, and Ca Concentrations and Final pH by 0.1 M Phosphate Depending on Lime Addition: GFH**

| Lime addition (g) | As(III) (μg/L) | As(V) (μg/L) | Total As (%) | Fe (mg/L) | Ca (mg/L) | Final pH |
|---|---|---|---|---|---|---|
| 0 | 150 | 14,800 | 11.2% | 3 | 1 | 9.17 |
| 1 | 321 | 21,400 | 16.2% | < 0.05 | 5 | 10.24 |
| 2 | 529 | 47,000 | 35.5% | < 0.05 | 2 | 11.80 |
| 3 | 422 | 41,300 | 31.1% | < 0.05 | 1 | 12.72 |
| 4 | 362 | 32,800 | 24.7% | < 0.05 | 1 | 12.95 |
| 5 | 210 | 20,300 | 15.3% | < 0.05 | 3 | 13.01 |
| 6 | < 1 | 5 | 0.004% | < 0.05 | 180 | 13.11 |
| 7 | < 1 | 1 | 0.001% | < 0.05 | 174 | 13.15 |
| 8 | < 1 | 1 | 0.001% | < 0.05 | 187 | 13.15 |
| 9 | < 1 | 1 | 0.001% | < 0.05 | 191 | 13.16 |
| 10 | < 1 | 1 | 0.001% | < 0.05 | 182 | 13.17 |

The leached arsenic concentrations compared to that of the zero lime (untreated GFH) increased when 1 to 5 g of lime was added. The behavior can be explained by the increased pH values of the treatments with increasing lime addition, which might have contributed to unfavorable conditions for As(V) adsorption. However, soluble arsenic concentrations in the leachate were reduced when greater than 5 g of lime was added. From the lime addition of 6 to 10 g, desorbed arsenic was not detected. At the range of 6 to 10 g of lime addition, excess soluble calcium, which did not join in any reactions, was extracted.

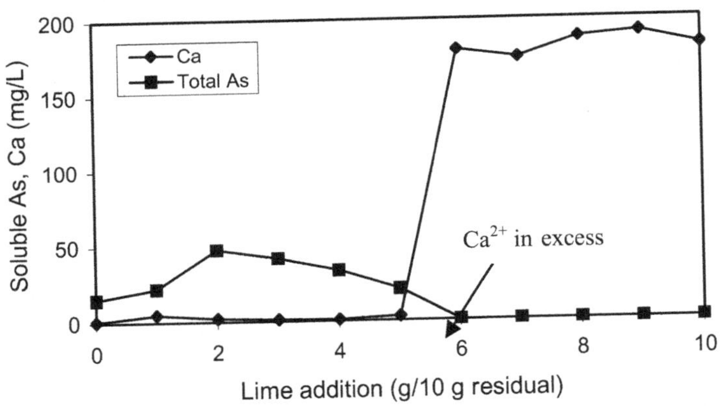


**Figure 6.1. Soluble Total As and Ca Concentrations by 0.1M Phosphate Extraction as a Function of Lime Addition: GFH**

There are three possible explanations for the results observed in Table 6.3. The first is that phosphate in the extraction solution was precipitated with the lime (Marani et al., 1997) or with calcite ($CaCO_3$) (Donnert and Salecker, 1999). In wastewater treatment, lime has been frequently used to remove phosphate. During the 7 day curing period, lime can react with atmospheric $CO_2$, and calcite ($CaCO_3$) generation might be possible. The pK values for phosphate are 2.2 for $pK_1$, 7.2 for $pK_2$, and 12.4 for $pK_3$. Thus, $PO_4^{3-}$ is the predominant species of phosphate at the high pH in the suspensions. $Ca_4H(PO_4)_3$ and /or $Ca_5(PO_4)_3OH$ might be generated as precipitated forms and this could be determined by X-ray diffraction analysis. At high pH, the release of As(V) and Fe should be substantial, but the results did not follow this expected trend. In brief, little or very low concentrations of As(V) and Fe were observed in the leachates even though the system pH was over 12. However, the removal of phosphate by calcium precipitation can not be discounted.

A second possible explanation is that calcium-iron-arsenic compounds were generated by calcium hydration during the 7-day curing period. The final pH of the leachates was very high, thus iron oxy-hydroxides should be dissolved as $Fe(OH)_4^-$. However, no soluble iron was detected. It is possible that calcium could play a critical role in preventing the iron leaching by the formation of calcium-iron compounds (Wilkie and Hering, 1996). Wilkie and Hering (1996) suggested the cooperative effect of calcium on the adsorption of As(V) at high pH. According to their description, the adsorption of $Ca^{2+}$ onto iron oxy-hydroxides resulted in a positive surface charge of the adsorbents in the high pH range. This reaction would result in favorable conditions for As(V) adsorption. The change of surface charge on Fe oxy-hydroxides in the presence or absence of calcium is illustrated in Figure 6.2.

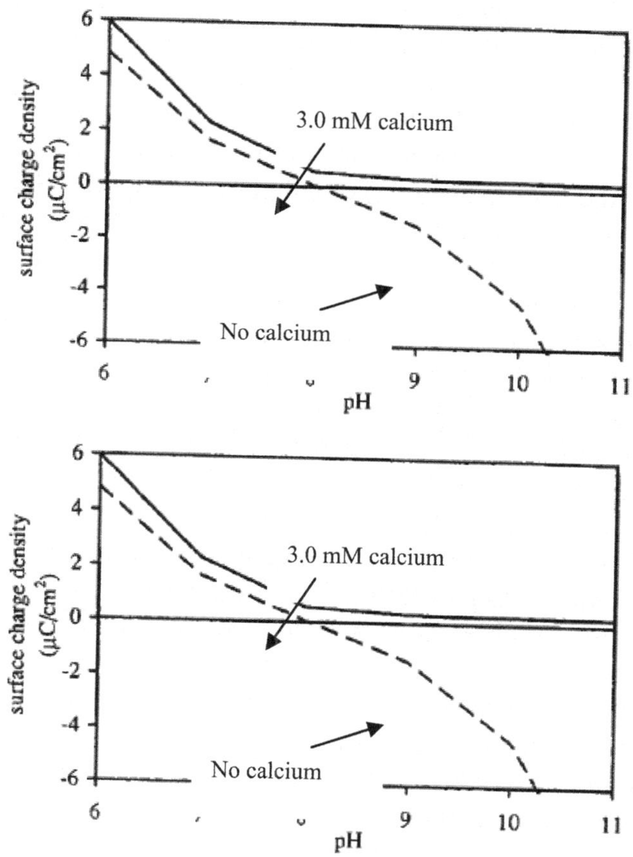


Source: Wilkie and Hering, 1996

**Figure 6.2. Surface Charge Density of Fe Hydroxide as Affected by Calcium: Model Predictions for TOT Fe = 50 μM in 0.01 M $NaNO_3$**

Cornwell et al. (2003) report that the presence of calcium substantially reduced arsenic release from ferric residuals, particularly at high pH. High pH was observed to result in arsenic release if calcium were not added. A similar effect was observed in the presence of magnesium, which supports the hypothesis that the mechanism for increased arsenic binding is one of counter ion adsorption. This mechanism assumes that calcium or magnesium is adsorbed to the surface, which increases surface charge. The increased surface charge increases the binding forces for negatively charged compounds such as arsenic. The suggested surface reactions for the complexation of Fe oxy-hydroxide and calcium and intrinsic adsorption constants (Schecher and McAvoy, 1994) are :

$$\equiv Fe^{w}OH + Ca^{2+} = \equiv Fe^{w}OCa^{+} + H^{+} \quad \text{Log } K^{int} = -5.85 \qquad (6.2)$$
$$\equiv Fe^{s}OH + Ca^{2+} = \equiv Fe^{s}OHCa^{2+} \quad \text{Log } K^{int} = 4.97 \qquad (6.3)$$

Enhanced adsorption of phosphate onto goethite from seawater has been reported when calcium is present and pH is high (Hawke et al., 1989). Identifying the mechanism for this effect in these experiments is difficult, because lime was added after GFH and arsenic were allowed to react for some time. It is not clear to what extent calcium added during stabilization experiments was able to fully react within 7 days reaction time. The chemistry of the phosphate extraction system for GFH residuals solidified by lime is complex. Therefore, a variety of experiments, analyses and supporting data are required in order to obtain an explanation.

A final explanation for the reduction of soluble arsenic concentration in the leachate is from the formation of less soluble calcium-arsenic compounds at high pH: calcium arsenate ($Ca_4(OH)_2(AsO_4)_2 \bullet 4H_2O$ and $Ca_4(AsO_4)_3OH$) or calcium arsenite ($CaHAsO_3$) (Kameswari et al., 2001; Bothe and Brown, 1999; Dutre and Vandecasteele, 1995). Solubility of iron hydroxides increases with the increase of system pH. Thus, the possibility of arsenic leaching was increased. Calcium can precipitate with the arsenic and form less soluble calcium-arsenic compounds, especially at high pH. This explanation is expected to be the most feasible one to interpret the data of Figure 6.1. Bothe and Brown (1999) studied the effects of phosphate on the possible release of As(V) solidified with lime. They reported the control of solution phase composition by arsenate apatite ($Ca_5(AsO_4)_3OH$) in the suspensions containing phosphate, the formation of arsenate apatite at the expense of phosphate apatite ($Ca_5(PO_4)_3OH$), and no observation of solid solution transition between $Ca_5(AsO_4)_3OH$ and $Ca_5(PO_4)_3OH$ under the ambient conditions. Cornwell et al. (2003) concluded that the effect of calcium on release of arsenic was most likely not due to formation of calcium arsenate solids. This conclusion was based on results of a batch experiment conducted with soluble calcium and arsenate at various pH. However, this experiment did not use any seed crystals of calcium arsenate, so the lack of arsenic removal from solution may be due to slow nucleation of the solid phase. In contrast, Jing et al. (2003), used FTIR analysis to conclude that addition of OPC to iron based water treatment residuals converts sorbed arsenate to a calcium arsenate solid phase that was only sparingly soluble. Zhu et al (2006) have reported values for the solubility product for calcium arsenate.

***Activated Alumina (AA)***

The S/S of AA with the addition of lime only was also carried out, and S/S procedures, curing, and extraction tests were applied as those to GFH. Tables 6.4 and 6.5 present the soluble As(III), As(V), Al, and Ca concentrations and final pHs in leachates by TCLP #2 extraction and 0.1 M phosphate extraction, respectively. In general, very low concentrations of As(III) and As(V) were leached by the TCLP as shown in Table 6.4. Therefore, the extraction method is not suitable to show the effects of lime addition for the reduction of arsenic leaching.

**Table 6.4. Extracted As, Al, and Ca Concentrations and Final pH by TCLP #2 Depending on Lime Addition: AA**

| Lime addition (g) | As(III) (μg/L) | As(V) (μg/L) | Total As (%) | Al (mg/L) | Ca (mg/L) | Final pH |
|---|---|---|---|---|---|---|
| 0 | 12 | 1 | 0.78% | 101 | 166 | 4.12 |
| 1 | 3 | 4 | 0.42% | 1 | 1,930 | 6.19 |
| 2 | < 1 | 3 | 0.18% | 10 | 1,780 | 10.20 |
| 3 | < 1 | 1 | <0.1% | 2 | 2,070 | 12.03 |
| 4 | < 1 | < 1 | <0.1% | 1 | 2,740 | 12.37 |
| 5 | < 1 | < 1 | <0.1% | 1 | 2,740 | 12.39 |
| 6 | < 1 | < 1 | <0.1% | 1 | 2,400 | 12.41 |
| 7 | < 1 | < 1 | <0.1% | 1 | 2,360 | 12.42 |
| 8 | < 1 | < 1 | <0.1% | 2 | 2,770 | 12.46 |
| 9 | < 1 | < 1 | <0.1% | 2 | 2,330 | 12.49 |
| 10 | < 1 | < 1 | <0.1% | 2 | 2,330 | 12.50 |

Upon phosphate extraction, As(V) was the predominant species of released arsenic due to the very high pH. Arsenic leaching increased as the amount of lime added was increased from 1 to 3, slowly reduced when 4 and 5 g of lime was added, and dropped rapidly when 6 and 7 g of lime was added. The arsenic leaching trend for AA was similar to that for GFH. The tendency of leached Al concentrations followed that of arsenic as shown in Table 6.5. This trend was different than that of leached Fe concentrations, which were not detected for GFH. The tendency of Ca concentrations detected in the leachates was opposite to that of arsenic. As expected, excessive amounts of Ca, which were not reacted during arsenic stabilization and remained dissolved in solution, were detected.

**Table 6.5. Extracted As, Al, and Ca Concentrations and Final pH by 0.1 M Phosphate Depending on Lime Addition: AA**

| Lime addition (g) | As(III) (μg/L) | As(V) (μg/L) | Total As (%) | Al (mg/L) | Ca (mg/L) | Final pH |
|---|---|---|---|---|---|---|
| 0 | 10 | 1,490 | 90.4% | 1 | 6 | 9.10 |
| 1 | 10 | 2,110 | 128% | 12 | 4 | 10.29 |
| 2 | 4 | 2,190 | 132% | 301 | 4 | 10.09 |
| 3 | 4 | 2,770 | 167% | 360 | 7 | 12.50 |
| 4 | 2 | 1,780 | 107% | 426 | 12 | 12.67 |
| 5 | 1 | 950 | 57.3% | 349 | 1 | 12.76 |
| 6 | < 1 | 93 | 5.60% | 65 | 10 | 12.81 |
| 7 | < 1 | < 1 | <0.2% | 4 | 148 | 12.87 |
| 8 | < 1 | 4 | 0.24% | 5 | 161 | 12.91 |
| 9 | < 1 | < 1 | <0.2% | 4 | 167 | 12.94 |
| 10 | < 1 | < 1 | <0.2% | 4 | 176 | 12.96 |

It is suggested that the main mechanism for the prevention of arsenic leaching from AA was the precipitation of arsenic compounds with lime, which generated stable calcium-arsenic compounds.

### Addition of OPC Only

#### *Granular Ferric Hydroxide (GFH)*

OPC is commonly employed as a binder for S/S. In this study, OPC was mixed at 1, 2, 3, and 5 g with 10 g of residual to determine the effects of OPC addition on arsenic leaching. The leached As(III), As(V), Fe, and Ca concentrations and final pH in leachates by TCLP and 0.1 M phosphate leaching tests are shown in Tables 6.6 and 6.7, respectively.

**Table 6.6. Extracted As, Fe, and Ca Concentrations and Final pH by TCLP #2 Depending on OPC Addition: GFH**

| OPC addition (g) | As(III) (μg/L) | As(V) (μg/L) | Total As (%) | Fe (mg/L) | Ca (mg/L) | Final pH |
|---|---|---|---|---|---|---|
| 0 | < 1 | 23 | 0.02% | 2 | 56 | 3.53 |
| 1 | < 1 | 7 | 0.01% | < 0.05 | 1,520 | 6.20 |
| 2 | < 1 | 13 | 0.01% | < 0.05 | 1,590 | 7.26 |
| 3 | < 1 | 32 | 0.02% | < 0.05 | 1,530 | 8.74 |
| 5 | < 1 | 17 | 0.01% | < 0.05 | 1,520 | 11.05 |

The arsenic concentrations detected in the leachates by the TCLP test were very low. The Ca concentrations were relatively constant. Therefore, it is thought that calcium might not play any role in arsenic immobilization due to the presence of very low soluble arsenic concentrations. A relationship between arsenic leaching and OPC addition was not evident.

In the phosphate leaching study, it was obvious that OPC addition did not have a positive influence on arsenic immobilization as shown in Table 6.7. The results might be explained by two reasons: the available amount of Ca and pH. About 20 to 25% of $Ca(OH)_2$ from OPC by weight is normally generated from cement hydration (Glasser, 1993). Therefore, the available amount of calcium, which can precipitate arsenic and form stable calcium-arsenic compounds, was lower when compared to the results of lime addition. Thus, the released arsenic concentrations were considerable. To overcome this problem, higher amounts of OPC would be required. As shown in Table 6.7, the final pH of leachates was high and this made for unfavorable conditions for As(V) adsorption. The relationship of total As concentration released and OPC addition is described in Figure 6.3.

**Table 6.7. Extracted As, Fe, and Ca Concentrations and Final pH by 0.1 M Phosphate Depending on OPC Addition: GFH**

| OPC addition (g) | As(III) (μg/L) | As(V) (μg/L) | Total As (%) | Fe (mg/L) | Ca (mg/L) | Final pH |
|---|---|---|---|---|---|---|
| 0 | 150 | 14,800 | 11.2% | 3 | 1 | 9.17 |
| 1 | 199 | 17,300 | 13.1% | < 0.05 | 4 | 10.10 |
| 2 | 226 | 25,000 | 18.8% | < 0.05 | 2 | 10.83 |
| 3 | 325 | 34,000 | 25.6% | < 0.05 | 1 | 11.71 |
| 5 | 297 | 32,500 | 24.5% | < 0.05 | 1 | 12.65 |

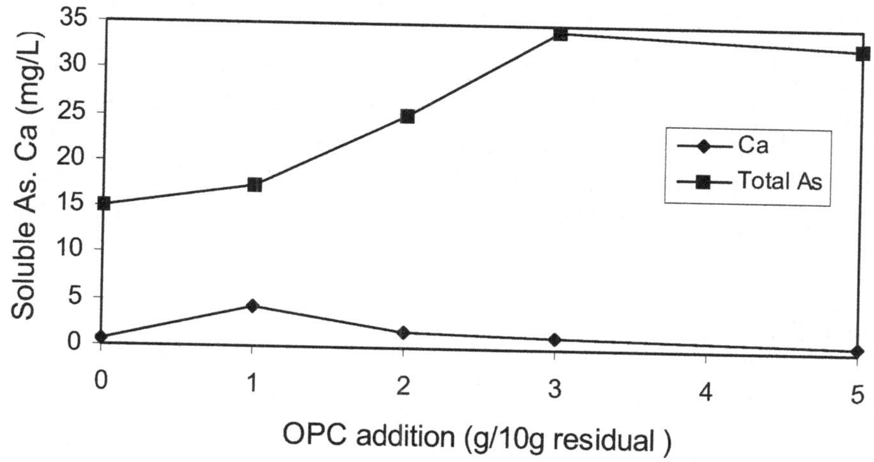


**Figure 6.3. Soluble total As and Ca Concentrations by 0.1 M Phosphate as a Function of OPC Addition: GFH**

*Activated Alumina (AA)*

The results of S/S with the addition of OPC only for AA are shown in Tables 6.8 and 6.9. The TCLP extraction test did not demonstrate any effects of OPC addition on arsenic leaching; the same as for GFH.

**Table 6.8. Extracted As, Al, and Ca Concentrations and Final pH by TCLP #2 Depending on OPC Addition: AA**

| OPC addition (g) | As(III) (μg/L) | As(V) (μg/L) | Total As (%) | Al (mg/L) | Ca (mg/L) | Final pH |
|---|---|---|---|---|---|---|
| 0 | 12 | 1 | 0.78% | 101 | 166 | 4.12 |
| 1 | 3 | 1 | 0.24% | 1 | 1,600 | 5.96 |
| 2 | 1 | 7 | 0.48% | 1 | 1,730 | 7.03 |
| 3 | < 1 | 11 | 0.66% | 1 | 1,780 | 10.25 |
| 5 | < 1 | 2 | 0.12% | 1 | 1,840 | 11.17 |

**Table 6.9. Extracted As, Al, and Ca Concentrations and Final pH by 0.1 M Phosphate Depending on OPC Addition: AA**

| OPC addition (g) | As(III) (μg/L) | As(V) (μg/L) | Total As (%) | Al (mg/L) | Ca (mg/L) | Final pH |
|---|---|---|---|---|---|---|
| 0 | 10 | 1,490 | 90.4% | 1 | 6 | 9.08 |
| 1 | 9 | 2,040 | 123% | 1 | 3 | 10.08 |
| 2 | 10 | 3,300 | 199% | 2 | 3 | 10.89 |
| 3 | 12 | 2,310 | 140% | 9 | 2 | 11.62 |
| 5 | 8 | 2,350 | 142% | 41 | 1 | 12.46 |

The main reason for the high soluble As(V) concentrations detected in the leachates is that the concentration of calcium in the system was not high enough. From the S/S study with the addition of OPC only to GFH and AA, it is suggested that OPC may not contribute positively to the immobilization of arsenic in residuals (Dutre and Vandecasteele, 1995). However, OPC could be used to improve the handling and the physical characteristics of the residuals (LaGrega et al, 1994).

## Addition of Lime and OPC

### *Granular Ferric Hydroxide (GFH)*

Six grams of lime and 3 and 5 g of OPC were selected based on the addition of lime only and OPC only studies and added to 10 g of residual. The results of TCLP and phosphate leaching tests are shown in Table 6.10. Two sets of binder-residual mixtures (6 g of lime + 3 g of OPC and 6 g of lime + 5 g of OPC) gave very similar results for arsenic leaching as expected.

**Table 6.10. Extracted As, Fe, and Ca Concentrations by TCLP #2 and 0.1 M Phosphate Extraction Depending on Lime and OPC Addition: GFH**

| TCLP | Lime (g) | OPC (g) | As(III) (μg/L) | As(V) (μg/L) | Total As (%) | Fe (mg/L) | Ca (mg/L) |
|---|---|---|---|---|---|---|---|
| | 6 | 3 | < 0.1 | 4 | <0.01% | 0.1 | 5,160 |
| | 6 | 5 | < 0.1 | 3 | <0.01% | 0.1 | 1,200 |
| | | | | | | | |
| **Phosphate** | **Lime (g)** | **OPC (g)** | **As(III) (μg/L)** | **As(V) (μg/L)** | **Total As (%)** | **Fe (mg/L)** | **Ca (mg/L)** |
| | 6 | 3 | < 0.1 | 18 | 0.01% | < 0.05 | 151 |
| | 6 | 5 | < 0.1 | 24 | 0.02% | < 0.05 | 127 |

When comparing the above results with those for the addition of lime only, there were no statistical differences for arsenic immobilization. However, the addition of lime only for the S/S of arsenic containing residuals poses a problem due to calcium-arsenic compounds decomposing slowly when exposed to atmospheric $CO_2$ (Riveros et al, 2001). The generally accepted equation for the decomposition reaction is

$$Ca_3(AsO_4)_2 + 3CO_2 + 3H_2O \rightarrow 3CaCO_3 + 2H_3ASO_4 \qquad (6.4)$$

Therefore, barriers which can block contact between the calcium-arsenic compounds and atmospheric $CO_2$ may be necessary for the management of arsenic immobilization in the residuals over a long period of time.

### *Activated Alumina (AA)*

Four sets of binder-residual mixtures were chosen based on the addition of lime only and OPC only studies. The results of the S/S with lime and OPC for AA are shown in Table 6.11.

From the above studies, it is suggested that the arsenic in the residuals can be successfully immobilized by the addition of lime and OPC. However, aging extraction tests for long time periods (e.g. over one year) are obviously necessary to demonstrate the efficiency of the stabilization of arsenic in water treatment residuals by lime and OPC addition.

**Table 6.11. Extracted As, Al, and Ca Concentrations by TCLP and Phosphate Extraction Depending on Lime and OPC Addition: AA**

| TCLP | Lime (g) | OPC (g) | As(III) (μg/L) | As(V) (μg/L) | Total As (%) | Al (mg/L) | Ca (mg/L) |
|---|---|---|---|---|---|---|---|
| | 6 | 3 | < 1 | < 1 | <0.2% | 2 | 2,090 |
| | 6 | 5 | < 1 | < 1 | <0.2% | 3 | 2,060 |
| | 7 | 3 | < 1 | < 1 | <0.2% | 2 | 2,140 |
| | 7 | 5 | < 1 | < 1 | <0.2% | 3 | 2,180 |
| Phosphate | Lime (g) | OPC (g) | As(III) (μg/L) | As(V) (μg/L) | Total As (%) | Al (mg/L) | Ca (mg/L) |
| | 6 | 3 | < 1 | 1 | <0.2% | 7 | 139 |
| | 6 | 5 | < 1 | 1 | <0.2% | 7 | 143 |
| | 7 | 3 | < 1 | 1 | <0.2% | 8 | 141 |
| | 7 | 5 | < 1 | < 1 | <0.2% | 6 | 150 |

## SUMMARY

The reduction of arsenic concentration in the leachate can be obtained by the addition of lime to the residuals due to the formation of less soluble and more stable calcium arsenic compounds. In practical aspects, solidified materials treated only by lime are not stable for the long term since the buffering effects of lime are decreased by dissolution and carbonation when exposed to atmospheric $CO_2$. Therefore, it is suggested that cement should be added with the lime to minimize the arsenic leaching over extended periods of time.

It is thought that the immobilization of arsenic in residuals can be achieved through the formation of less soluble calcium arsenic compounds by using lime and cement. Therefore, solidification and stabilization (S/S) techniques with lime and cement can be successfully applied for the immobilization of elevated arsenic concentrations in water treatment residuals.

# CHAPTER 7: CONCLUSIONS AND RESEARCH NEEDS

## CONCLUSIONS

The following conclusions are based on the range of conditions used in the experiments presented in this report.

- Residuals based on lime-soda ash have lower water contents, but bind arsenic less strongly and release it more easily than residuals based on aluminum or ferric iron.
- Residuals based on aluminum are less effective in removing arsenic from solution than those based on iron and are less effective in resisting release of arsenic during leaching experiments.
- Arsenate, compared to arsenite, is more effectively removed by residuals based on iron and aluminum and it is more strongly bound to the residuals.
- Addition of clays (kaolinite, montmorillonite) did not improve binding of arsenic to residuals.
- The amount of arsenite released from the residuals under conditions of the TCLP increased with time and reached a relatively constant value after about 10 hours. The amount of arsenate leached decreased slightly over time.
- Phosphate was effective in increasing the amount of arsenate released from synthetic and water treatment plant residuals. The concentrations used in these studies (10 mM, 100 mM) were much higher than would be found in natural water supplies, but their use did provide a method for comparing the strength with which arsenic was bound to different residuals. The extent to which arsenic would be released at lower concentrations of phosphate is not known.
- Sulfate and chloride had little to no effect on arsenic release under the conditions studied. The effects of anions on release of arsenic from aluminum residuals were difficult to measure, due the generally high amounts released without addition of anions.
- Arsenic release is affected by pH. Extreme pH (pH 4, pH 10) resulted in higher releases of arsenic unless calcium was added. Higher releases of arsenic were usually accompanied by higher releases of the primary element of the residual (aluminum or iron), indicating that the structure of the oxy-hydroxides was degrading.
- Residuals can have a an effect on pH. Aluminum and iron residuals tend to raise or lower pH in the direction of pH 7, while lime-soda residuals tend to raise pH under all conditions studied. A surface complex model can describe these effects.
- Presence of calcium helps reduce release of arsenic from residuals. This effect is observed when the residuals are produced in the presence of added calcium or when calcium is added to residuals in the form of lime as part of a S/S treatment. This impact of calcium was observed in both the presence and absence of phosphate.
- S/S treatment with lime or OPC can be successfully applied to the immobilization of elevated arsenic concentrations in water treatment residuals.

## RESEARCH NEEDS

Potential environmental impacts of disposal of arsenic-contaminated water treatment plant residuals are due primarily to the release of arsenic. Most residuals consist primarily of iron or aluminum oxy-hydroxides. The nature of binding between these surfaces and arsenic determines the extent to which arsenic will be retained on the solid phases after disposal. The current study has clearly demonstrated the advantages of iron-based versus aluminum- and calcium-based residuals. The current study has also demonstrated the benefits of calcium in the production and stabilization of water treatment residuals. More studies are needed at the pilot plant level. The influence of reduced conditions and alternating oxidized and reduced conditions on the release of arsenic needs to be more thoroughly evaluated for both iron and aluminum hydroxide systems and especially for mixed Fe/Al and mixed Fe/Al/Ca systems.

# APPENDIX A: ARSENIC ANALYSIS

## INTRODUCTION

Arsenic analytical techniques for quantification as described in the Standard Methods (APHA et al, 1998) are

1) Electrothermal Atomic Absorption Spectrometry – Standard Method 3113B
2) Hydride Generation/Atomic Absorption Spectrometry (HG-AAS) – Standard Method 3114B and 3114C
3) Inductively Coupled Plasma (ICP) Method – Standard Method 3120B
4) Inductively Coupled Plasma/Mass Spectrometry (ICP/MS) Method – Standard Method 3125B

Thus, furnace atomic absorption spectrometry (FAAS), HG-AAS, ICP, or ICP/MS is required to analyze arsenic concentration in waters. Affordable instruments in most water treatment plants and laboratories are FAAS and HG-AAS when comparing the prices of equipment and supplies. In this section, arsenic analysis by hydride generation with atomic absorption spectrometry (HG-AAS) is described in detail. In particular, the three standard curve method for arsenic speciation is illustrated.

## ARSENIC ANALYSIS BY HYDRIDE GENERATION

The inorganic arsenic forms (As(V) and As(III)) must be converted (reduced) to arsine ($AsH_3$) to be analyzed by HG-AAS. A mixture of sodium borohydride ($NaBH_4$) and HCl is employed as reducing agents to generate $AsH_3$ from As(V) or As(III) (Hering and Elimelech, 1996; Korte and Fernando, 1991; Schmidt and Royer, 1973).

As(V) is first reduced to As(III), followed by transformation to $AsH_3$ (called two-electron reduction), but As(III) can be directly reduced to $AsH_3$. Therefore, the reduction rate of As(V) is slower than that of As(III). This difference in kinetic reactions is more pronounced at high pH. Thus, reduction rate is slower at high pH than at low pH. By controlling the pH of the reaction solution, inorganic arsenic speciation can be achieved (Aggett and Aspell, 1976; Braman and Foreback, 1973).

From the difference of reduction kinetics of As(III) versus As(V), two different approaches can be employed for arsenic analysis in As(III) and As(V) mixture systems. First, As(V) is pre-reduced with a pre-reductant such as potassium iodide (KI) or L-cysteine to overcome the incomplete generation of $AsH_3$ from As(V) by $NaBH_4$. However, the concentrations of both As(III) and As(V) cannot be determined by this approach. The second approach is "the three standard curve method" that can amend the difference of reduction kinetics and finally allows for the speciation of As(III) and As(V)

## THREE STANDARD CURVE METHOD FOR ARSENIC SPECIATION

The three standard curve method for inorganic arsenic speciation was prepared and developed by Loeppert and Biswas (Unpublished). The description of the theory and the procedure is summarized in this section.

**Theory**

The analytical absorbance (A) of a mixture of As(III) and As(V) by HG-AAS is equal to the absorbance attributable to each As(III) and As(V). Thus, at pH 6.5 or pH 1.0 can be expressed by:

$$A_{6.5} = A_{As(III),pH6.5} + A_{As(V),pH6.5} \quad (A.1)$$

$$A_{1.0} = A_{As(III),pH1.0} + A_{As(V),pH1.0} \quad (A.2)$$

The slope of the standard curve is written by:

$$\text{Slope} = \frac{\text{absorbance}}{\text{concentration}} = \frac{A}{C} \quad (A.3)$$

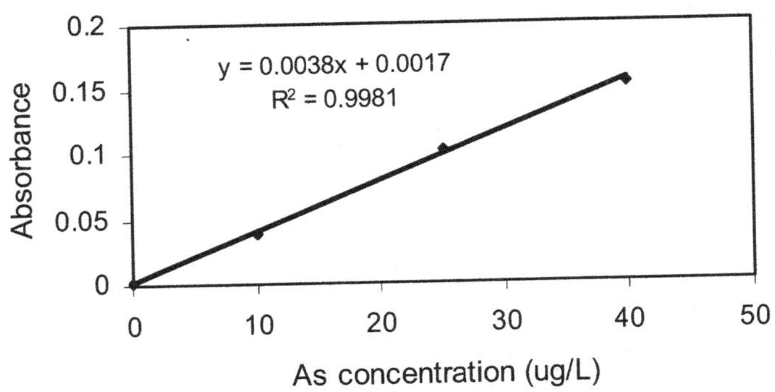


**Figure A.1. Example for the Relationship Between Concentration and Absorbance**

Figure A.1 is one example which shows the relationship of standard concentration versus absorbance. Equation A.3 can be substituted into equations A.1 and A.2 and arsine formation from As(V) is zero at pH 6.5. Therefore, equations A.1 and A.2 can be simplified to,

$$A_{6.5} = aC_{As(III)} \quad (A.4)$$

$$A_{1.0} = cC_{As(III)} + dC_{As(V)} \quad (A.5)$$

where, a = slope of As(III) standard curve at pH 6.5
c = slope of As(III) standard curve at pH 1.0
d = slope of As(V) standard curve at pH 1.0

Equations A.4 and A.5 can be further modified to:

$$C_{As(III)} = \frac{A_{6.5}}{a} \quad (A.6)$$

$$C_{As(V)} = \frac{\{[A_{1.0} - c(C_{As(III)})\}}{d} \quad (A.7)$$

Each As(III) and As(V) concentration can be determined by Equations A.6 and A.7, respectively, and total As concentration is the sum of As(III) and As(V) concentrations.

**Procedure**

Primary As(III) and As(V) stock solutions (1 g/L) should be prepared from $As_2O_3$ and $As_2O_5$ (or $Na_2HAsO_4 \bullet H_2O$), respectively. The detailed preparation procedure is in Standard Method 3114B (APHA et al, 1998). The secondary stock solution (10 mg/L) is prepared by the dilution of the primary stock solution. Analytical standard solutions are prepared using the secondary stock solutions. The concentrations can be 0 (blank), 5, 10, 20, 40 μg/L. The reagents used are summarized in Table A.1. Concentration of each reagent can be altered depending on the analytical instrument. The sample/acid, reductant, and carrier gas flow rates should be determined by preliminary tests for each specific analytical instrument.

**Table A.1. Reagents used for arsenic analysis**

| Parameter | Value |
|---|---|
| Borohydride concentration | 1.5% (m/v) $NaBH_4$ in 0.5% NaOH |
| Phosphate buffer | 0.2 M $NaH_2PO_4$ (monobasic), pH 3.0 adjusted by phosphoric acid |
| HCl concentration | 5.0 M |

Analysis with the mixture of borohydride solution and phosphate buffer corresponds to the reaction of pH 6.5 and the reaction from mixture of borohydride and HCl solutions is referred to the reaction of pH 1.0. It is possible to accurately analyze the As(III), As(V), and total As concentration from the three standard method. First, (i) As(III) standards and samples should be analyzed with the mixture of borohydride solution and phosphate buffer. From the As(III) standard calibration curve, the concentration of each sample can be calculated using the equation A.6. Subsequently, (ii) As(III) and As(V) standards and samples (the same as those used in (i)) are subject to analyze with a mixture of borohydride and HCl solutions. The two calibration curves for As(III) and As(V) standards are obtained. Using the equation A.7, As(V) concentration of each sample can be also determined. Total As concentration is the sum of As(III) and As(V) concentrations. Basically, three standard curves, As(III) standard curve obtained from analysis with a mixture of borohydride solution and phosphate buffer, and As(III) and As(V) standard curves obtained from analysis with a mixture of borohydride and HCl solutions, will be gained from this analysis.

# APPENDIX B: TOXICITY CHARACTERISTIC LEACHING PROCEDURE

## SUMMARY OF METHOD

The solid phase is extracted with an amount of extraction fluid equal to 20 times the weight of the solid phase. Following extraction, the liquid extract is separated from the solid phase by filtration through a 0.2 μm membrane filter.

The TCLP test determines if a particular waste meets the applicable technology-based treatment standards to be land-disposed and is also widely used to evaluate the effectiveness of stabilization.

## APPARATUS AND MATERIALS

Apparatus and materials used include:

1) Rotary agitation apparatus
2) Filtration devices
3) Filters (0.2 μm pore size)
4) pH meter
5) Laboratory balance
6) Beaker and volumetric flask – 500 mL
7) Magnetic stirrer
8) Capped polypropylene bottles – 200 mL
9) 9.5 mm standard sieve

## REAGENTS

Reagents used include the following:

1) DI water – water purification system (Millipore Super-Q or equivalent)
2) Hydrochloric acid (1N), HCl – ACS reagent grade
3) Nitric acid (1N), $HNO_3$ – ACS reagent grade
4) Sodium hydroxide (1N), NaOH – ACS reagent grade
5) Glacial acetic acid, $CH_3CH_2OOH$ – ACS reagent grade
6) Hydroperoxide, $H_2O_2$ – ACS reagent grade
7) Extraction fluid #1 (0.10 M acetic acid and 0.064 M NaOH) – Add 5.7 mL glacial $CH_3CH_2OOH$ to 500 mL of reagent water, add 64,3 mL of 1 N NaOH, and dilute to a volume of 1 L. When correctly prepared, the pH of this fluid will be 4.93±0.05.
8) Extraction fluid #2 (0.10 M acetic acid) – Dilute 5.7 mL glacial $CH_3CH_2COOH$ with DI water to a volume 1 L. When correctly prepared, the pH of this fluid will be 2.88±0.05.
9) Analytical standards

## PROCEDURE

**Table B.1. Test variables**

| | |
|---|---|
| Leachant-to-waste ratio | Liquid-to-solid weight ratio of 20:1 (200 mL: 10 g) |
| Surface area of the waste | Particle size smaller than 9.5 mm. |
| Type of leachant | Weak acetic acid extraction liquid (0.10 M acetic acid) |
| pH of the leachant | pH 4.93 and pH 2.88 |
| Contact time | 0, 1, 2, 4, 8, 18, 36, 72 hr |
| Extent of agitation | 30±5 rpm |
| Extraction vessel | Polyethylene bottle |
| Temperature | Room temperature (22±3°C) |
| Number of replacements of fresh leachant | No |

The dewatered residual samples are crushed to a particle size smaller than 9.5 mm (check using a 9.5 mm standard sieve).

The crushed sample is mixed with the extraction fluid #1, in a liquid-to-solid weight ratio of 20:1 (200 mL of liquid: 10 g of sample). After that, close the bottle tightly using a cap. Mix well and collect the sample around 15 mL to know the initial concentrations of As(III), As(V), Fe, Al, and Ca.

Place the sample and liquid mixed bottles into a rotary agitation apparatus and conduct agitation at room temperature.

Aliquots (around 15 mL) of suspension samples are collected at desirable time intervals (1, 2, 4, 8, 18, 36, 72 hr).

After extraction, the sample is filtered to separate liquid and solids with a 0.2 μm membrane filter. The filtrate is defined as the TCLP extract. Measure pH of the extract and record it.

All extracts must be kept under refrigeration (4°C) until analyzed.

The extracts are further prepared and analyzed according to the analytical methods. Measure arsenic speciation (As(III) and As(V)) using HG-AAS and Fe, Al, and Ca concentrations are measured using FAAS.

# APPENDIX C: PAINT FILTER TEST

## INTRODUCTION

USEPA method 9095 was used to conduct the Paint Filter Test. The disposal of liquid hazardous wastes in a sanitary landfill is banned and should not contain free liquids (RCRA Subtitle D). The paint filter test is adequate to evaluate the presence of free liquids in materials prior to disposal in a landfill.

## SUMMARY OF METHOD

A predetermined amount of material is placed in a paint filter. If any portion of the material passes through and drops from the filter within the 5-min test period, the material is deemed to contain free liquids.

## APPARATUS AND MATERIALS

Apparatus and materials used include the following:

1) Conical paint filter: Mesh number 60±5% (fine meshed size). Available at local paint stores such as Sherwin-Williams and Glidden.
2) Glass funnel: If the paint filter cannot sustain its weight on the ring stand, then a fluted glass funnel or glass funnel with a mouth large enough to allow at least 1 in. of the filter mesh to protrude should be used to support the filter.
3) Ring stand and ring, or tripod.
4) Graduated cylinder or beaker: 100-mL.

## SAMPLE COLLECTION, PRESERVATION, AND HANDLING

A 100-g representative sample is required for the test.

## PROCEDURE

Assemble test apparatus as shown in Figure C.1. Place sample (100-g of dewatered residual) in the filter. The particles to be tested should be reduced smaller than 1 cm (i.e., should be capable of passing through a 9.5 mm (0.375 inch) standard sieve.) Grinding sorbent materials should be avoided as this may destroy the integrity of the sorbent and produce many "fine particles" which would normally not be present.

For brittle materials larger than 1 cm that do not conform to the filter, light crushing to reduce oversize particles is acceptable if it is not practical to cut the material. Materials such as clay, silica gel, and some polymers may fall into this category.

Allow the sample to drain for 5 min into the graduated cylinder.

If any portion of the test material collects in the graduated cylinder in the 5-min period, then the material is deemed to contain free liquids.

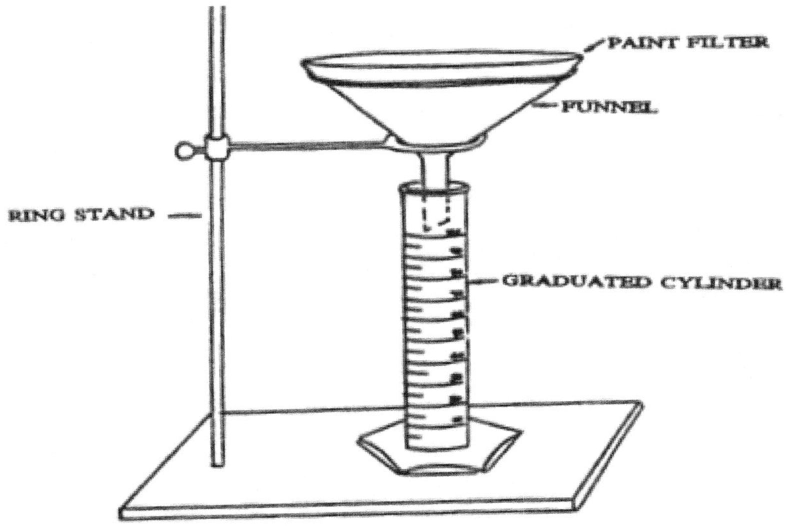


**Figure C.1. Apparatus for Paint Filter Test**

| Sample Name | Free Liquid Presence |
|---|---|
| A | Yes, or No |
| | |

# APPENDIX D: TOTAL, FIXED, AND VOLATILE SOLIDS IN SOLID

Standard method 2540G was used to measure total, fixed and volatile solids in solid samples (APHA, AWWA and WEF, 1998)

## SUMMARY OF THE METHOD

A well-mixed residuals sample is placed in a weighing dish and dried to constant weight in an oven at a temperature of 103 to 105°C. Volatile solids are determined by placing the sample in a furnace and heating to a temperature of 550±50°C. The remaining solids represented the fixed solids (inorganic contents), while the weight lost is the volatile solids (organic contents).

## APPARATUS

The following were used in the test procedure:

1) Evaporating dishes: Dishes of 100-mL capacity
2) Muffle furnace: capable of operation at 550°C
3) Steam bath
4) Desiccator
5) Drying oven: capable of operation at 103 to 105°C
6) Analytical balance: capable of weighing to 10 mg (0.01 g)

## PROCEDURE

### Total solids:

Preparation of evaporating dish – If volatile solids are to be measured, ignite a clean evaporating dish at 550°C for 1 hr in a muffle furnace. If only total solids are to be measured, heat dish at 103 to 105°C for 1 hr in an oven. Cool in desiccator, weigh, and store in desiccator until ready for use.

For fluid samples the following steps are followed:

1) If the sample contains enough moisture to flow more or less readily, stir to homogenize, place 25 to 50 g in a prepared evaporating dish, and weigh.
2) Evaporate to dryness on a water bath, dry at 103 to 105°C for 1 hr, cool to balance temperature in an individual desiccator containing fresh desiccant, and weigh.
3) Repeat heating, cooling, desiccating, and weighing procedure until the weight change is less than 4% or 50 mg, whichever is less.
4) Analyze at least 10% of all samples in duplicate. Duplicate determinations should agree within 5% of their average weight.

For solid samples, the following steps are followed:

1) If the sample consists of discrete piece of solid material (e.g., dewatered sludge), take cores from each piece with a No. 7 cork borer or pulverize the entire sample coarsely on a clean surface by hand, using rubber gloves.

2) Place 25 to 50 g in a prepared evaporating dish and weigh.
3) Place in an oven at 103 to 105°C overnight (for 24 hr).
4) Cool to balance temperature in a desiccator for at least 30 min and weigh.
5) Repeat drying (for 1 hr), cooling, weighing, and desiccating steps until weight change is less than 4% or 50 mg.
6) Analyze at least 10% of all samples in duplicate. Duplicate determination should agree within 5% of their average weight.

**Fixed and volatile solids:**

Transfer the dried residue to a cool muffle furnace, heat furnace to 550°C, and ignite for 1 hr. Cool in desiccator to balance temperature for 30 min and weigh.

Repeat igniting (30 min), cooling, desiccating and weighing steps until the weight change is less than 4% or 50 mg.

Analyze at least 10% of all samples in duplicate. Duplicate determination should agree within 5% of their average weight.

**Calculations**

$$\%\ \text{total solids} = \frac{(A - B) \times 100}{C - B}$$

$$\%\ \text{volatile solids} = \frac{(A - D) \times 100}{A - B}$$

$$\%\ \text{fixed solids} = \frac{(D - B) \times 100}{A - B}$$

where:

A = weight of dried residue + dish. (g)

B = weight of dish. (g)

C = weight of wet sample + dish. (g)

D = weight of residue + dish after ignition. (g)

| Sample Name | Total solids (%) | Volatile solids (%) | Fixed solids (%) |
|---|---|---|---|
| | | | |

# APPENDIX E: ACID DIGESTION

USEPA Method 3050B was followed to conduct acid digestions.

## SUMMARY OF THE METHOD

For the digestion of samples, 1 g (dry weight) sample is digested with repeated additions of nitric acid ($HNO_3$) and hydrogen peroxide ($H_2O_2$)

For flame atomic absorption spectrometry (FAAS) analyses, hydrochloric acid (HCl) is added to the initial digestate and the sample is refluxed. The digestate is filtered and the filter paper and residues are rinsed, first with hot HCl and then hot reagent water. Filter paper and residue are returned to the digestion flask, refluxed with additional HCl and then filtered again. The digestate is then diluted to a final volume of 100 mL.

Aluminum, calcium, copper, iron, magnesium, manganese, sodium, and zinc concentrations are analyzed using FLAA and total arsenic concentration is analyzed using HG-AAS.

## APPARATUS AND MATERIALS

Digestion vessels – 250 mL

Vapor recovery device. (e.g., ribbed watch glasses, appropriate refluxing device, appropriate solvent handling system)

Drying ovens – able to maintain 30±4°C.

Filter paper – Whatman No. 41 or equivalent.

Analytical balance – capable of accurate weighing to 0.01 g.

Heating source – adjustable and able to maintain a temperature of 90 to 95°C. (e.g., hot plate, block digestor, etc)

Funnel or equivalent.

Volumetric flasks – 100 mL

## REAGENTS

Reagent water – DI water

Nitric acid (concentrated), $HNO_3$.

Hydrochloric acid (concentrated), HCl.

Hydrogen peroxide (30%), $H_2O_2$.

## PROCEDURE

Mix the sample thoroughly to achieve homogeneity. For each digestion, weigh to the nearest 0.001 g and transfer 1 g sample (dry weight) to a digestion vessel.

Add 10 mL of 1:1 $HNO_3$, mix the slurry, and cover with a watch glass or vapor recovery device.

Heat the sample to 95±5°C and reflux for 10 to 15 minutes without boiling.

Allow the sample to cool, add 5 mL of concentrated $HNO_3$, replace the cover, and reflux for 30 minutes.

If brown fumes are generated, indicating oxidation of the sample by $HNO_3$, repeat this step (addition of 5 mL of conc. $HNO_3$) over and over until no brown fumes are given off by the sample indicating the complete reaction with $HNO_3$.

Using a ribbed watch glass or vapor recovery system, either allow the solution to evaporate to approximately 5 mL without boiling or heat at 95±5°C without boiling for two hours. Maintain a covering of solution over the bottom of the vessel at all times.

After completing the above steps, the sample has cooled for 5 minutes, add slowly 10 mL of 30% $H_2O_2$. Care must be taken to ensure that losses do not occur due to excessive vigorous effervescence.

Continue to add 30% $H_2O_2$ in 1-mL aliquots with warming until the effervescence is minimal or until the general sample appearance is unchanged. Do not add more than a total of 10 mL 30% $H_2O_2$.

Cover the sample with a ribbed watch glass or vapor recovery device and continue heating the acid-peroxide digestate until the volume has been reduced to approximately 5 mL or heat at 95±5°C without boiling for two hours.

Add 10 mL conc. HCl to the sample digest and cover with a watch glass or vapor recovery device. Place the sample on/in the heating source and reflux at 95±5°C for 15 minutes.

Filter the digestate through Whatman No. 41 filter paper and collect filtrate in a 100-mL volumetric flask.

The diluted digestate solution contains approximately 5% (v/v) $HNO_3$. For analysis, withdraw aliquots of appropriate volume and add any required reagent or matrix modifier.

## CALCULATIONS

The concentrations determined are to be reported on the basis of the actual weight of the sample. If a dry weight analysis is desired, then the percent solids of the sample must also be provided.

If percent solids is desired, a separate determination of percent solids must be performed on a homogeneous aliquot of the sample.

| Sample | Fe (mg/kg) | Al (mg/kg) | - | - | - |
|---|---|---|---|---|---|
| | | | | | |
| | | | | | |

# APPENDIX F: PH MEASUREMENT

The technique for pH measurement was taken from those developed for soils (Soil pH and Soil Activity, *Methods of Soil Analysis*, Soil Science Society of America 1996)

## SUMMARY OF METHOD

The sample is mixed with reagent water, and the pH of the resulting aqueous solution is measured

## APPARATUS AND MATERIALS

pH meter equipped with glass and reference electrode, or combination electrode.
Beakers – 100 mL
Pipet or automatic pipet of 10 mL.
Analytical balance: capable of weighing 0.1 g

## REAGENTS

Deionized water.
Standard buffers, pH 4, pH 7, and pH 10.
0.01 M $CaCl_2$ solution.
1 M KCl solution.

## PROCEDURE

Determination – pH in water
Weigh out 10 g of air-dry waste in a 100-mL beaker.

Add 10 mL of deionized water to the waste in the beaker and continuously stir the suspension for 5 min. A stirring stick, or stirring machine can be sued but care should be taken to minimize contamination. (For large-scale determinations, a shaking machine can be employed as is done in most soil testing laboratories.)

Let the waste suspension stand for about 15 min to allow most of the suspended waste to settle out from the suspension.

Swirl the suspension in the beaker and insert the electrodes into the suspension. Electrodes may be placed in the clear supernatant above the soil, directly in the sedimented sludge, or the entire suspension may be stirred during the pH determination. The important thing is that the measurements be carried out in a consistent way. In water pH determination, values taken in the supernatant generally will be slightly higher than in the stirred suspension. With a salt pH, the differences between the three techniques practically disappear. For exposed glass electrodes, it is useful to have a stop of some type so that the bulb will not contact the bottom of the beaker.

Read pH and record as $pH_w$.

Electrodes must be thoroughly rinsed between samples (between pH readings).

pH in One-One Hundredth Molar Calcium Chloride (0.01 M $CaCl_2$)

a. Repeat the above procedure (a-d) but use 0.01 M $CaCl_2$ instead of water to make the soil suspension.

Read pH and record as $pH_{CaCl2}$.

pH in One Molar Potassium Chloride

a. Repeat steps *a* to *d* step but use 1 M KCl instead of water to make the sludge suspension.

b. Read pH and record as $pH_{KCl}$.

# APPENDIX G: CATION EXCHANGE CAPACITY

USEPA Method 9081 was used to determine cation exchange capacity.

## SUMMARY OF METHOD

The sludge sample is mixed with an excess of sodium acetate solution, resulting in an exchange of the added sodium cations for the matrix cations. Subsequently, the sample is washed with isopropyl alcohol. An ammonium acetate solution is then added, which replaces the adsorbed sodium with ammonium. The concentration of displaced sodium is then determined by atomic absorption.

## APPARATUS AND MATERIALS

The following were used:

1) Centrifuge tube and stopper: 50-mL, round-bottom, narrow neck.
2) Mechanical shaker.
3) Volumetric flask: 100-mL.

## REAGENTS

Sodium acetate (NaAc), 1.0 N – If needed, add a few drops of acetic acid or NaOH solution to bring the solution to pH 8.2.

Ammonium acetate ($NH_4Ac$), 1.0 N – If needed, add more $NH_4OH$ to obtain a pH of 7

Isopropyl alcohol: 99%

## PROCEDURE

The following steps were followed:

1) Weigh 4 g of medium particle size of sample and transfer the sample to a 50-mL, round-neck centrifuge tube. (particle size is less than 0.425 mm by grinding)
2) Add 33 mL of 1.0 N sodium acetate solution, stopper the tube, shake it in a mechanical shaker for 5 min, and centrifuge it until the supernatant liquid is clear.
3) Decant the liquid, and repeat 4.2. step three more times.
4) Add 33 mL of 99% isopropyl alcohol, stopper the tube, shake it in a mechanical shaker for 5 min, and centrifuge it until the supernatant liquid is clear.
5) Repeat the procedure described in 4.4. step two more times.
6) Add 33 mL of ammonium acetate solution, stopper the tube, shake it in a mechanical shaker for 5 min, and centrifuge it until the supernatant liquid is clear. Decant the washing into a 100-mL volumetric flask.
7) Repeat the procedure described in 4.6. step two more times.
8) Dilute the combined washing to the 100-mL mark with ammonium acetate solution and determine the sodium concentration by flame atomic absorption (FLAA) method.

## CALCULATIONS

The CEC of a residual is defined as the sum of the exchangeable cations of that residual expressed in milliequivalents for 100 g of dry residual (meq/100g = cmol/kg). A high CEC generally denotes a potentially high capacity to retain contaminants in the residual.

## REFERENCES


Akhter, H., Cartledge, F.K., Roy, A., and Tittlebaum, M.E. 1997. Solidification/Stabilization of Arsenic Salts: Effects of Long Cure Times. *J. Hazard. Mater.* 52:2-3:247.

Akhter H., Cartledge F. K., Miller J., and McLearn M. Treatment of arsenic-contaminated soils. II: Treatability study and remediation. *Journal of Environmental Engineering* 2000, 126(11), 1004-1012.

Aggett, J., and Aspell, A.C. 1976. Determination of Arsenic(III) and Total Arsenic by Atomic-Absorption Spectroscopy. *Analyst.* 101(1202):341-347.

Alcan Chemicals. 1998. Product Data – Activated Aluminas. Division of Alcan Aluminum Corporation. Cleveland, OH.

APHA, AWWA, and WEF (American Public Health Association, American Water Works Association, and Water Environment Federation). 1998. *Standard Methods for the Examination of Water and Wastewater.* 20th ed. Washington, D.C.

Amy, G., Edwards, M., Benjamin, M., Carlson, K., Chwirka, J., Brandhuber, P., McNeil, L., and Vagliasindi, F. 2000. Arsenic Treatability Options and Evaluation of Residuals Management Issues. Denver, CO: AWWARF and AWWA.

Anderson, L.C.D., and Bruland, K.W. 1991. Biogeochemistry of Arsenic in Natural-Waters – The Importance of Methylated Species. *Environ. Sci. Technol.* 25(3):420-427.

Anderson, M.A., Ferguson, J.F., and Gavis, J. 1976. Arsenate Adsorption on Amorphous Aluminum Hydroxide. *J. Colloid Interf. Sci.* 54(3)391-399.

Andreae, M.O. 1981. Arsenic: Industrial, Biomedical, Environmental Perspectives. *In: Proceedings of the Arsenic Symposium.* Lederer, W.H. and Fensterheim, R.J. (Eds.). Van Nostrand, New York. pp. 378-391.

Artiola, J.F., Zaiback, D., and Johnson, S.H. 1990. In Situ Treatment of Arsenic Contaminated Soil from a Hazardous Industrial Site: Laboratory Studies. *Waste Manage.* 10:73-78.

Baes, C.F., Mesmer, R. E. (1976). *The Hydrolysis of Cations.* Krieger Publishing Company. 358-365.

Bates, M.N., Smith, A.H., and Hopenhayn-Rich, C. 1992. Arsenic Ingestion and Internal Cancers: A Review. *Am. J. Epidemiol.* 135(5):462-476.

Benefield, L.D., and Morgan, J.S. 1990. Chemical Precipitation. *In: Water Quality and Treatment.* Edited by F.W. Pontius. 4th ed, New York: McGraw-Hill, pp. 641-708.

Berg, M., Tran, H.C., Nguyen, T.C., Pham, H.V., Schertenleib, R., and Giger, W. 2001. Arsenic Contamination of Groundwater and Drinking Water in Vietnam: A Human Health Threat. *Environ. Sci. Technol.* 35(13):2621-2626.

Bose, P., and Sharma, A. 2002. Role of Iron in Controlling Speciation and Mobilization of Arsenic in Subsurface Environment. *Water Res.* 36(19):4916-4926.

Bothe, J.V.Jr., and Brown, P.W. 1999. Arsenic Immobilization by Calcium Arsenate Formation. *Environ. Sci. Technol.* 33(21):3806-3811.

Braman, R.S., and Foreback, C.C. 1973. Methylated Forms of Arsenic in Environment. *Science* 182(4118):1247-1249.

Carbonell-Barrachina, A.A., Jugsujinda, A., Burlo, F., Delaune, R.D., and Patrick, W.H. 1999. Arsenic Chemistry in Municipal Sewage Sludge as Affected by Redox Potential and pH. *Wat. Res.* 34(1):216-224.

Chakraborti, D., Biswas, B.K., Chowdhury, T.R., Basu, G.K., Mandal, B.K., Chowdhury, U.K., Mukherjee, S.C., Gupta, J.P., Chowdhury, S.R., Rathore, K.C. 1999. Arsenic Groundwater Contamination and Sufferings of People in Rajnandgaon district, Madhya Pradesh, India. *Curr. Sci. India* 77(4):502-504.

Cheng, R.C., Liang, S., Wang, H.C., and Beuhler, M.D. 1994. Enhanced Coagulation for Arsenic Removal. *J. Am. Water Works Assoc.* 86(9):79-90.

Cherry, J.A., Shaikh, A.U., Tallman, D.E., and Nicholson, R.V. 1979. Arsenic Species as an Indicator of Redox Conditions in Groundwater. *J. Hydrol.* 43(1-4):373-392.

Conner, J.R. 1993. Chemistry of Cementitious Solidified/Stabilized Waste Forms. *In: Chemistry and Microstructure of Solidified Waste Forms*. Edited by R.D. Spence. Boca Raton, FL: Lewis Publishers, pp. 41-82.

Cornwell, D.A., MacPhee, M.J., Mutter, R., Novak, J., Edwards, M., 2003. Disposal of Waste Resulting From Arsenic Removal Processes, AWWARF and AWWA.

Cox, C.D., Ghosh, M.M. "Surface Complexation of Methylated Arsenates by Hydrous Oxides." Water Resources, 28(5): 1181-1188.

Cullen, W.R., and Reimer, K.J. 1989. Arsenic Speciation in the Environment. *Chem. Rev.*, 89(4):713-764.

Davis, M.L., and Cornwell, D.A. 1998. Softening. *In: Introduction to Environmental Engineering*. New York: McGraw-Hill, pp. 178-199.

Donnert, D., and Salecker, M. 1999. Elimination of Phosphorous from Waste Water by Crystallization. *Environ. Technol.* 20(7):735-742.

Driehaus, W., Jekel, M., and Hildebrandt, U. 1998. Granular Ferric Hydroxide – A New Adsorbent for the Removal of Arsenic from Natural Water. *J. Water SRT – Aqua* 47(1):30-35.

Dutre, V., and Vandecasteele, C. 1995. Solidification/Stabilization of Hazardous Arsenic Containing Waste from a Copper Refining Process. *J. Hazard. Mater.* 40(1):55-68.

Dutre, V., and Vandecasteele, C. 1998. Immobilization Mechanism of Arsenic in Waste Solidified Using Cement and Lime. *Environ. Sci. Technol.* 32(18):2782-2787.

Edwards, M. 1994. Chemistry of Arsenic Removal During Coagulation and Fe-Mn Oxidation. *J. Am. Water Works Assoc.* 86(9):64-78.

Ferguson, J.F., and Gavis, J. 1972. A Review of the Arsenic Cycle in Natural Waters. *Wat. Res.* 6(11):1259-1274.

Fox, K. R. (1989). "Field Experience with Point-of-Use Treatment Systems for Arsenic Removal." *Journal AWWA,* 81(2): 94.

Frank, P., Clifford, D. (1986). "Arsenic(III) Oxidation and Removal from Drinking Water," EPA-600/S2-86/021, U. S. EPA, Cincinnati, OH.

Fuessle, R.W., and Taylor, M.A. 2000. Stabilization of Arsenic- and Barium-Rich Glass Manufacturing Waste. *J. Environ. Eng.-ASCE* 126(3):272-278.

Ghosh, M. M., Yuan, J. R. (1987). "Adsorption of Inorganic Arsenic and Organoarsenicals on Hydrous Oxides." *Environmental Progress*, 6(3): 15

Glasser, F.P. 1993. Chemistry of Cement-Solidified Waste Forms. *In: Chemistry and Microstructure of Solidified Waste Forms.* Edited by R.D. Spence. Boca Raton, FL: Lewis Publishers, pp. 1-39.

Gulledge, J.H., and O'Connor, J.T. 1973. Removal of Arsenic(V) from Water by Adsorption on Aluminum and Ferric Hydroxides. *J. Am. Water Works Assoc.* 65(8):548-552.

Gupta, S. K., Chen, K.Y. (1978). "Arsenic Removal by Adsorption." *Journal WPFC*: 493.

Hawke, D., Carpenter, P.D., and Hunter, K.A. 1989. Competitive Adsorption of Phosphate on Goethite in Marine Electrolytes. *Environ. Sci. Technol.* 23(2):187-191.

Hering, J.G., and Elimelech, M. 1996. Arsenic Removal by Enhanced Coagulation and Membrane Processes. Denver, CO: AWWARF and AWWA.

Hering, J.G., Chen, P., Wilkie, J.A., Elimelech, M., and Liang, S. 1996. Arsenic Removal by Ferric Chloride. *J. Am. Water Works Assoc.* 88(4):155-167.

Hiemstra, T., and Van Riemsdijk, W.H. 1999. Surface Structural Ion Adsorption Modeling of Competitive Binding of Oxyanions by Metal (Hydr)oxides. *J. Colloid Interf. Sci.* 210(1):182-193.

Hingston, F.J., Posner, A.M., and Quirk, J.P. 1971. Competitive Adsorption of Negatively-charged Ligands on Oxide Surfaces. *Discuss. Faraday Soc.* 52:334-342.

Hongshao, Z., and Stanforth, R. 2001. Competitive Adsorption of Phosphate and Arsenate on Goethite. *Environ. Sci. Technol.* 35(24):4753-4757.

Hooper, K., Iskander, M., Sivia, G., Hussein, F., Hsu, J., Deguzman, M., Odion, Z., Ilejay, Z., Sy, F., Petreas, M., and Simmons, B. 1998. Toxicity Characteristic Leaching Procedure Fails to Extract Oxoanion-Forming Elements That Are Extracted by Municipal Solid Waste Leachates. *Environ. Sci. Technol.* 32(23):3825-3830.

Hsieh, H.N., and Raghu, D. 1997. Criteria Development for Water Treatment Plant Residual Monofills. Denver, CO: AWWARF and AWWA.

Jackson, B.P., and Miller, W.P. 2000. Effectiveness of Phosphate and Hydroxide for Desorption of Arsenic and Selenium Species from Iron Oxides. *Soil Sci. Soc. Am. J.* 64(5):1616-1622.

Jain, A., and Loeppert, R.H. 2000. Effect of Competing Anions on the Adsorption of Arsenate and Arsenite by Ferrihydrite. *J. Environ. Qual.* 29(1):1422-1430.

Jain, C.K., and Ali, I. 2000. Arsenic: Occurrence, Toxicity and Speciation Techniques. *Wat. Res.* 34(17):4304-4312.

Jing, C., Korfiatis, G.P., Meng, X. 2003. Immobilization Mechanisms of Arsenate in Iron Hydroxide Sludge Stabilized with Cement. *Environ. Sci. Technol.* 37(21):5050-5056.

Kameswari, K.S.B., Bhole, A.G., and Paramasivam, R. 2001. Evaluation of Solidification/Stabilization (S/S) Process for the Disposal of Arsenic-Bearing Sludges in Landfill Sites. *Environ. Eng. Sci.* 18(3):167-176.

Kim, J., Davis, A.P., and Kim, K. 2003. Stabilization of Available Arsenic in Highly Contaminated Mine Tailings Using Iron. *Environ. Sci. Technol.* 37(1):189-195.

Knowles, F.C., and Benson, A.A. 1983. The Biochemistry of Arsenic. *Trends Biochem. Sci.* 8(5):178-180.

Korte, N.E., and Fernando, Q. 1991. A Review of Arsenic(III) in Groundwater. *Crit. Rev. Env. Contr.* 21(1):1-39.

Kuhlmeier, P.D., Sherwood, S.P. (1996). "Treatability of Inorganic Arsenic and Organoarsenicals in Groundwater." *Water Environment Research,* 68(5): 946.

LaGrega, M.D., Buckingham, P.L., and Evans, J.C. 1994. Stabilization and Solidification. *In: Hazardous Waste Management*. New York: McGraw-Hill, pp. 641-704.

Langmuir, D. 1997a. Aqueous Complexes. *In: Aqueous Environmental Geochemistry*. Upper Saddle River, NJ: Prentice-Hall, pp. 82-122.

Langmuir, D. 1997b. Chemical Weathering. *In: Aqueous Environmental Geochemistry*. Upper Saddle River, NJ: Prentice-Hall, pp. 231-265.

Lehmann, M., Zouboulis, A.I., Matis, K.A. (1999). "Removal of Metal Ions from Dilute Aqueous Solutions: A Comparative Study of Inorganic Sorbent Material." *Chemosphere,* 39(6): 881-892.

Lepkowski, W. 1998. Arsenic Crisis in Bangladesh. *Chem. Eng. News* 76(46):27-29.

Loeppert, R.H., and Biswas, B.K. Unpublished. Arsenic Analysis by Hydride Generation. In Methods of Analysis for Soil Arsenic. Texas A&M University.

Loeppert, R.H., Jain, A., El-Haleem, M.A.A., and Biswas, B.K. 2003. Quantity and Speciation of Arsenic in Soils by Chemical Extraction. *In: Biogeochemistry of Environmentally Important Trace Elements*, ACS Sym. Ser. 835:42-56.

Marani, D., Di Pinto, A.C., Ramadori, R., and Tomei, M.C. 1997. Phosphate Removal from Municipal Wastewater with Low Lime Dosage. *Environ. Technol*. 18(2):225-230.

Mariner, R.H., and Willey, L.M. 1976. Geochemistry of Thermal Waters in Long-Valley, Mono-County, California. *J. Geophys. Res.* 81(5):792-800.

Masscheleyn, P.H., Delaune, R.D., and Patric, W.H.Jr. 1991. Effect of Redox Potential and pH on Arsenic Speciation and Solubility in a Contaminated Soil. *Environ. Sci. Technol.* 25(8):1414-1419.

Meng, X., Korfiatis, G.P., Jing, C., and Christodoulatos, C. 2001. Redox Transformations of Arsenic and Iron in Water Treatment Sludge during Aging and TCLP Extraction. *Environ. Sci. Technol.* 35(17):3476-3481.

Masscheleyn, P.H., Pardue, J.H., DeLaune, R.D., Patrick, Jr., W.H. (1992). "Chromium Redox Chemistry in a Lower Mississippi Valley BottomLand Hardwood Wetland." *Environmental Science and Technology,* 26(6): 1217-1226.

McNeill, L.S., Edwards, M. (1995). "Soluble Arsenic Removal at Water Treatment Plants." *Journal AWWA,* 87(4): 105.

McNeill, L.S., Edwards, M. (1997) "Arsenic Removal During Precipitative Softening." *Journal of Environmental Engineering ASCE*, 123(5): 453-460

Miller, J.P. 1996. *In-Situ Solidification/Stabilization of Arsenic Contaminated Soils*. Electric Power Research Institute, Report TR-106700. Palo Alto, CA.

Moskowicz P, Brna R, Mehu J, van der Sloot H, Hoede D. "Leaching behavior assessment of waste solidified with hydraulic binders: critical study of diffusional approach. In: Environmental aspects of construction with waste materials (Goumans Ed.) JJJM. Amsterdam: Elsevier Science, 1994, 421-431.

Nishimura T, Ito Ch. T, Tozawa K, and Robins R. G. "The calcium-arsenic-water-air system." Proceedings of the 15$^{th}$ Hydrometallurgical Meeting, Vancouver(Canada), 1985, 1-19.

Nickson, R., McArthur, J., Burgess, W., Ahmed, K.M., Ravenscroft, P., and Rahman, M. 1998. Arsenic Poisoning of Bangladesh Groundwater. *Nature* 395:338-338.

Nriagu, J., Pacnya, J. (1988). "Quantitative Assessment of Worldwide Contamination of Air, Water, and Soils by Trace Metals." *Nature,* 333: 134-139.

Palfy, P., Vircikova, E., and Molnar, L. 1999. Processing of Arsenic Waste by Precipitation and Solidification. *Waste Manage*. 19(1):55-59.

Peterson, M.L., and Carpenter, R. 1983. Biogeochemical Processes Affecting Total Arsenic and Arsenic Species Distributions in an Intermittently Anoxic Fjord. *Mar. Chem.* 12(4):295-321.

Pontius, F.W., Brown, K.G., Chen, C. (1994). "Health Implications of Arsenic in Drinking Water." *Journal AWWA*, 86(9): 52.

Raven, K.P., Jain, A., and Loeppert, R.H. 1998. Arsenite and Arsenate Adsorption on Ferrihydrite: Kinetics, Equilibrium, and Adsorption Envelops. *Environ. Sci. Technol.* 32(3):344-349.

Richard, F.C., Bourg, A.C.M. (1991). "Aqueous Geochemistry of Chromium: A Review." *Water Research*, 25(7): 807-816.

Riveros, P.A., Dutrizac, J.E., and Spencer, P. 2001. Arsenic Disposal Practices in the Metallurgical Industry. *Can. Metall. Quart.* 40(4):395-420.

Roy, A., Eaton, H.C., Cartledge, F.K., and Tittlebaum, M.E. 1992. The Effect of Sodium Sulfate on Solidification/Stabilization of a Synthetic Electroplating Sludge in Cementitious Binders. *J. Hazard. Mater.* 30(3):297-316.

Rubel, F., Hathaway, S.W. (1985). "Pilot Study for Removal of Arsenic from Drinking Water at the Fallon, Nevada, Naval Air Station," EPA-600/S2-B5/094, U.S. EPA, Cincinnati, OH.

Sadiq, M. 1997. Arsenic Chemistry in Soils: An Overview of Thermodynamic Predictions and Field Observations. *Water Air Soil Poll.* 93(1-4):117-136.

Sandesara, M.D. 1978. Process for Disposal of Arsenic Salts. U.S. Patent 4, 118, 243.

Sanyal, S.K., DeDatta, S.K. (1991). "Chemistry of Phosphorus Transformation in Soils." Advanced Soil Science, 16: 1-20.

Sax, N.I. (1984). *Dangerous Properties of Industrial Materials*, 6th ed., Van Nostrand Reinhold, New York, 178.

Schecher, M.D., and McAvoy, D.C. 1994. MINEQL$^+$ v. 3.01, Users Manual, Hallowell, ME.

Schindler, P W. (1981). Surface Complexes at Oxide-Water Interfaces. Ann Arbor, Michigan, Ann Arbor Science Publishers, Inc.

Schmidt, F.J., and Royer, J.L. 1973. Sub Microgram Determination of Arsenic, Selenium, Antimony and Bismuth by Atomic-Absorption Utilizing Sodium-borohydride Reduction. *Anal. Lett.* 6(1):17-23.

Schultz, M.F., Benjamin, M.M., Ferguson, J. (1987). "Adsorption and Desorption of Metals on Ferrihydrite: Reversibility of the Reaction and Sorption Properties of the Regenerated Solid." *American Chemical Society*, 21(9): 863-868.

Simms J., Upton, J., and Barnes, J. 2000. Arsenic Removal Studies and the Design of a 20,000 $m^3$ Per Day Plant in the UK. AWWA Inorganic Contaminants Workshop, Albuquerque, NM, February 27-29, 2000.

Smedley, P.L., and Kinniburgh, D.G. 2002. A review of the sourcee, behavior and distributionof arsenic in natural waters. *Applied Geochemistry* 17: 517-568.

Smith, J.D. 1973. Arsenic, Antimony and Bismuth. *In: Comprehensive Inorganic Chemistry*, Volume 2. Pergamon Press. pp. 580-683.

Smith, A.H., Hopenhayn-Rich, C., Bates, M.N., Goeden, H.M., Hertzpicciotto, I., Duggan, H.M., Wood, R., Kosnett, M.J., and Smith, M.T. 1992. Cancer Risks from Arsenic in Drinking-Water. *Environ. Heath Persp.* 97:259-267.

Stumm, W., Morgan, J.J. (1970). Aquatic Chemistry: An Introduction Emphasizing Chemical Equilibria in Natural Waters. Wiley-Interscience.

Tamaki, S., and Frankenberger, W.T.Jr. 1992. Environmental Biochemistry of Arsenic. *Rev. Environ. Contam. T.* 124:79-110.

Taylor, H.F.W. 1997. Cement Chemistry, 2nd ed. London, England: Thomas Telford.

United States Environmental Protection Agency. 1990. Final Best Demonstrated Available Technology (BDAT) Background Document for K031, K084, K101, K102, Characteristic As Waste (D004), Characteristic Se Wastes (D010), and P and U Wastes Containing As and Se Listing Constituents, EPA-530-SW-90-059A.

United States Environmental Protection Agency .1992. Test Methods for Evaluating Solid Waste, Physical/Chemical Methods – SW846. 3rd ed. U.S. Government Printing Office: Washington, DC.

United States Environmental Protection Agency. 1998 *Special Report on Ingested Inorganic Arsenic*. Skin Cancer; Nutritional Essentiality, EPA/625/3-87/013, Washington, D. C.

United States Environmental Protection Agency. 1998. Superfund Remediation at Rocker Timber Framing and Treating Plant near Butte, MT. Region 8, Denver, CO.

United States Environmental Protection Agency. 2000. National Primary Drinking Water Regulations; Arsenic and Clarifications to Compliance and New Source Contaminants Monitoring; Proposed Rule. Federal Register, 65:38888.

United States Environmental Protection Agency, Office of Water. 2000 "Technical Fact Sheet: Proposed Rule for Arsenic in Drinking Water and Clarifications to Compliance and New Source Contaminants Monitoring [EPA 815-F-00-011]," Available: prop_techfs.htmL

United States Environmental Protection Agency. 2001. National Primary Drinking Water Regulations; Arsenic and Clarifications to Compliance and New Source Contaminants Monitoring; Final Rule. Federal Register, 66:6975.

United States Environmental Protection Agency. 2001 "EPA to Propose Withdrawal of Arsenic in Drinking Water Standard; Seeks Independent Reviews," Press Release

United States Environmental Protection Agency. 2002. Arsenic Treatment Technologies for Soil, Waste, and Water. Solid Waste and Emergency Response. EPA-542-R-02-004.

United States Geological Survey (2001). Bauxite and Alumina Statistics and Information. http://minerals.usgs.gov/minerals/pubs/commodity/bauxite

Voigt, D.E., Brantley, S.L., and Hennet, R.J.C. 1996. Chemical Fixation of Arsenic in Contaminated Soils. *Appl. Geochem*. 11(5):633-643.

Welch, A.H., Lico, M.S., and Hughes, J.L. 1988. Arsenic in Ground Water of the Western United States. *Ground Water* 26(3):333-347.

Wilkie, J.A., and Hering, J.G. 1996. Adsorption of Arsenic onto Hydrous Ferric Oxide: Effect of Adsorbate/Adsorbent Ratios and Co-occurring Solutes. *Colloid Surface A*. 107:97-110.

Williams, M., Fordyce, F., Paijitprapapon, A., and Charoenchaisri, P. 1996. Arsenic Contamination in Surface Drainage and Groundwater in Part of the Southeast Asian Tin Belt, Nakhon Si Thammarat Province, Southern Thailand. *Environ. Geol.* 27(1):16-33.

World Health Organization. (1998) "Water, Sanitation and Health: Guidelines for Drinking Water Quality," Guidelines for Drinking-Water Quality, 2nd ed., Vol. 1 (Geneva: World Health Organization, 1993) p. 41-42 Available: int/water_sanitation_ health/GDWQ/Chemicals/arsenics.htm

Yaziz, M.I., Chin, F.L., Tang, S.N., and Bich, N.N. 1999. Heavy Metal Leaching of Solidified Sludge from a Glass Components Industry. *J. Environ. Sci. Health, A*. 34(4):853-861.

Zhu, Y.N., Zhang, X.H., Xie, Q.L., Wang, D.Q., Cheng, G.W. 2006. Solublity and Stability of Calcium Arsenates at 25 °C. *Water, Air and Soil Pollution* 169: 221-238.

# ABBREVIATIONS

| | |
|---|---|
| $\beta$ | Buffer Capacity |
| $\mu g$ | Microgram |
| $\mu m$ | Micrometer, Micron |
| $\equiv$ | Surface Specie |
| $\equiv AlO^-$ | Negatively Charged Surface Aluminum Hydroxide |
| $\equiv AlOH$ | Uncharged Surface Aluminum Hydroxide |
| $\equiv AlOH_2^+$ | Positively Charged Surface Aluminum Hydroxide |
| $^\circ C$ | Degrees Celsius |
| AA | Activated Alumina |
| AAS | Atomic Absorption Spectrometry |
| AC | Activated Carbon |
| Al | Aluminum |
| $Al_2O_3$ | Aluminum Oxide, Alumina |
| $Al_2SO_4$ | Aluminum Sulfate, Alum |
| As | Arsenic |
| $A_s$ | Specific Surface Area, Area per Unit Mass |
| As(III) | Trivalent Arsenic, Arsenite |
| As(V) | Pentavalent Arsenic, Arsenate |
| $As_2S_3$ | Arsenic Sulfide (Orpiment) |
| $AsH_3$ | Arsine |
| AsS | Arsenic Sulfide (Realgar) |
| BDL | Below Detection Limit |
| C | Total Concentration of Acid and All of Its Conjugate Bases |
| Ca | Calcium |
| $C_a$ | Concentration of Acid |
| $Ca(OH)_2$ | Calcium Hydroxide, Hydrated Lime |
| $CaCO_3$ | Calcium Carbonate |
| CaO | Calcium Oxide, Quicklime |

| | |
|---|---|
| $C_b$ | Concentration of Base |
| Cl | Chloride |
| $CO_2$ | Carbon Dioxide |
| C-S-H | Calcium Silicate Hydrate |
| D | Dose of Sorbent |
| DI | De-ionized |
| EBCT | Empty Bed Contact Time |
| Eh | Electrode Potential |
| FAAS | Flame Atomic Absorption Spectrometry |
| Fe | Iron |
| $FeCl_3$ | Ferric Chloride |
| g | Gram |
| GFH | Granular Ferric Hydroxide |
| GFO | Granular Ferric Oxide |
| $H_2O_2$ | Hydrogen Peroxide |
| $H_3AsO_3$ | Arsenous Acid |
| $H_3AsO_4$ | Arsenic Acid |
| HG-AAS | Hydride Generation Atomic Absorption Spectrometry |
| $HNO_3$ | Nitric Acid |
| hr | Hour |
| ICP | Inductively Coupled Plasma |
| ICP-MS | Inductively Coupled Plasma-Mass Spectrometry |
| IOCS | Iron Oxide-Coated Sand |
| $K^-$ | Surface Equilibrium Constant for Second Deprotonation |
| $K^+$ | Surface Equilibrium Constant for First Deprotonation |
| Kaol | Kaolinite |
| kg | Kilogram |
| $K_{sp}$ | Solubility Product |
| L | Liter |

| | |
|---|---|
| M | Molar, Mol/Liter |
| MCL | Maximum Contaminant Level |
| MF | Microfiltration |
| mg | Milligram |
| Mg | Magnesium |
| mL | Milliliter |
| mM | Millimolar |
| mm | Millimeter |
| Mn | Manganese |
| Mont | Montmorillonite |
| Na | Sodium |
| $Na_2CO_3$ | Sodium Carbonate, Soda Ash |
| $Na_2H_2PO_4$ | Sodium Di-hydrogen Phosphate |
| $Na_2SO_4$ | Sodium Sulfate |
| $NaBH_4$ | Sodium Borohydride |
| $NaCH_3COO$ | Sodium Acetate |
| NaCl | Sodium Chloride |
| $NaHCO_3$ | Sodium Bicarbonate |
| $NaNO_3$ | Sodium Nitrate |
| NaOH | Sodium Hydroxide |
| NASF | Naval Air Station Fallon |
| NF | Nanofiltration |
| $N_t$ | Total Number of Surface Sites per Unit Area |
| OPC | Ordinary Portland Cement |
| Org | Organic |
| pE | Negative Log of Electron Activity |
| $pK_a$ | Negative Log of Acid Disassociation Constant |
| ppb | Parts Per Billion |
| Q | Concentration of Surface Charge per Unit Mass |

| | |
|---|---|
| RO | Reverse Osmosis |
| S/S | Solidification/Stabilization |
| TCLP | Toxicity Characteristic Leaching Procedure |
| TOT≡AlOH | Total Concentration of Aluminum Hydroxides |
| UF | Ultrafiltration |
| USEPA | United States Environmental Protection Agency |
| WTP | Water Treatment Plant |

IWA Publishing's authorised EU representative for General Product Safety Regulations is Diane D'Arras, 15 rue Duret, 75116 Paris, France, e-mail: safety@iwap.co.uk.

Printed and bound by CPI Group (UK) Ltd, Croydon, CR0 4YY
17/09/2026
02218798-0012